RECREATION ON THE COLORADO RIVER

by

Dirksen and Reeves

Sail Sales Publishing
P.O. Box 1028
Aptos, California 95001
Phone: 408-662-2456

CREDITS

MAPS by RENEE REEVES

CARTOONS by GREG DIRKSEN

TYPESETTING by NANCY GORHAM

PRINTING by DELTA LITHOGRAPH COMPANY

COPYRIGHT © 1985

First Edition

ISBN: 0-943798-07-8

Cover Photo: Pittsburg Point, Lake Havasu

Every effort has been made to assure the accuracy of this publication. The authors, publisher and their dealers are not responsible for factors beyond their control.

ORDER FORM

SEND TO: Sail Sales Publishing
P.O. Box 1028
Aptos, CA 95001

☐ "Recreation Lakes of California" — Sixth Edition

   $9.95   Book
    .65   Tax
   1.00   Postage & Handling
   $11.60 CHECK ENCLOSED

☐ "Recreation on the Colorado River" — First Edition

   $9.95   Book
    .65   Tax
   1.25   Postage & Handling
   $11.85 CHECK ENCLOSED

☐ Both Publications (Free Postage & Handling)

   $19.90 Books
   1.29   Tax
   $21.19 CHECK ENCLOSED

NAME: _____

ADDRESS: _____

_____

NOTES

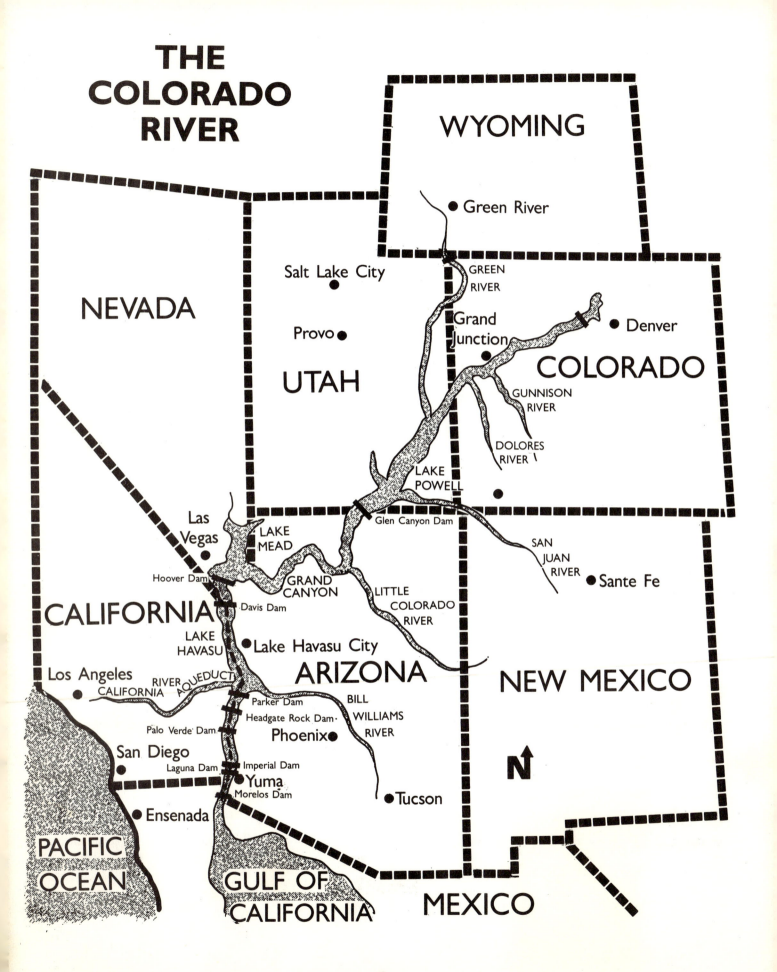

THE COLORADO RIVER

WYOMING

• Green River

NEVADA

Salt Lake City •

GREEN
RIVER

Grand
Junction •

• Denver

Provo •

UTAH

COLORADO

GUNNISON
RIVER

DOLORES
RIVER

LAKE
POWELL

Las
Vegas •

LAKE
MEAD

Glen Canyon Dam

SAN
JUAN
RIVER

• Sante Fe

Hoover Dam

CALIFORNIA

GRAND
CANYON

LITTLE
COLORADO
RIVER

• Davis Dam

LAKE
HAVASU

• Lake Havasu City

Los Angeles •

RIVER AQUEDUCT

ARIZONA

NEW MEXICO

CALIFORNIA

Parker Dam

BILL
WILLIAMS
RIVER

Headgate Rock Dam •

Palo Verde Dam •

Phoenix •

San Diego •

Laguna Dam

Imperial Dam

• Yuma

Morelos Dam

• Tucson

• Ensenada

N

PACIFIC
OCEAN

GULF OF
CALIFORNIA

MEXICO

INTRODUCTION

After many years of continuous research, extensive travel and a lot of fun, we are happy to present the First Edition of RECREATION ON THE COLORADO RIVER. Using the same format as our Bestseller, RECREATION LAKES OF CALIFORNIA, this Guide defines in a clear, concise manner, the many attractions and facilities of this wonderful River.

The Colorado River is one of America's, if not the world's, most spectacular scenic and recreational assets. Originating high in the Colorado Rockies, this mighty River carves its way through the geological splendor of the Western Slope of the United States through Canyonlands, Glen Canyon and the Grand Canyon into the desert lands of Arizona, Nevada, California and Mexico. Dropping over two miles in its 1,400 mile journey to the Sea of Cortez, the environment and mood of the River is ever changing.

Beginning as a quiet trickle of melted snow near Milner Pass on the Continental Divide, the Colorado's waters provide an unsurpassed source of recreational abundance. Whitewater enthusiasts rave about the challenges found in Westwater, Cataract and Grand Canyons. A canoe trip down the lower Colorado provides a meditative tour of nature's wealth. Sail in the world's highest elevation regatta at Grand Lake, houseboat amid the scenic splendor of Lake Powell, waterski on Lake Mead, or enjoy the boating excitement of "Parker Strip." The angler will find huge trout at Lee's Ferry, big stripers in Lake Havasu, and enormous catfish below Blythe. There is also a variety of trout, bass, crappie and other gamefish to be found along the River's course. Hunting, birding, nature study, Indian artifacts, swimming, hiking and backpacking are but a few of the many opportunities on or near the River. Recreation on the Colorado is for everyone who enjoys the outdoors.

Information is always changing. Camping, marine and recreation facilities are being developed, expanded and improved. The conditions of the River itself are subject to the whims of nature and man. Facilities are sometimes flooded or stranded by low water. These changes are our constant concern so that we may give you a current description of the River's resources.

We are ever thankful to the many people who have contributed to this Book with their support and timely suggestions. Local, State and Federal Agencies were indispensible in helping us compile the many details for this publication. The helpful people who operate the many facilities and the various Chambers of Commerce were a pleasure to work with. Most of all, we want to thank you, our readers, who have given us the inspiration and encouragement to complete RECREATION ON THE COLORADO RIVER. This is your book. Please continue your support by sending us your suggestions.

CONTENTS

This book is divided into four sections as the Colorado River flows through Colorado, Utah, Arizona, Nevada, California and Mexico.

Each section of the book is marked with a black line at the bottom of every page for easy identification.

Sections

COLORADO	Page 9 to 20
UTAH	Page 21 to 44
ARIZONA/NEVADA	Page 45 to 67
ARIZONA/CALIFORNIA/MEXICO	Page 68 to 92

General Information

No Trace Camping	Page 22
Hypothermia & Drinking Water	Page 56
Hot Weather Exploring	Page 88
Governmental Agencies	Pages 93, 94, 95
River Running	Page 96
Commercial Outfitters	Page 97, 98
Rapids Graph	Page 99
Rapid Ratings	Page 100
Pets	Page 101
Endangered Fish	Page 102, 103
INDEX	Pages 104, 105, 106, 107

Map Symbols

▲ Campground

△ Primitive/Undeveloped Campground

▲▲ Group Campground

■ Picnic Site

▬ Launch Ramp

▭ Primitive/Undeveloped Launch Ramp

♠ Resort

★ Marina

🚻 Toilets

P Parking

--- Graded Road

······· Four-Wheel Drive Road

⛳ Golf Course

🏠 Ranger Station

✈ Airport

— --- — State Border

▨ River or Lake

● Town or City

⬟ United States Interstate Highway

⬭ United States Federal Highway

○ State Highway

N⬆ North

These maps are not to scale.

Please do not use them for navigational purposes.

COLORADO

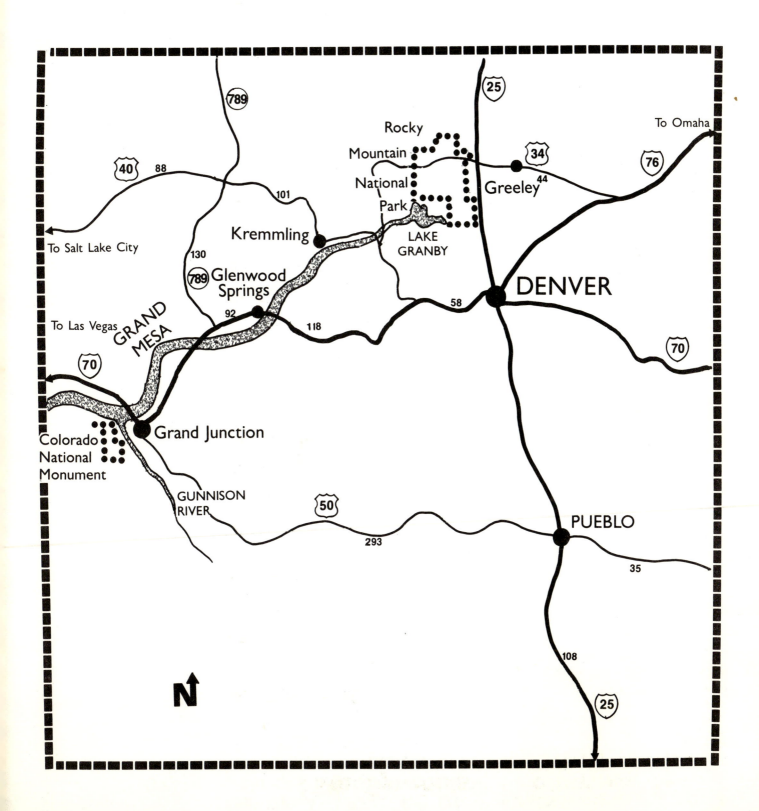

THE GREAT LAKES OF THE ROCKIES
GRAND, SHADOW MOUNTAIN, GRANBY, MONARCH &
WILLOW CREEK RESERVOIR

The Colorado River begins its majestic journey in this area of spectacular beauty and abundant recreational opportunities. Originating in the lofty peaks of the Rocky Mountain National Park straddling the Continental Divide, the Colorado River begins its western flow into the "Great Lakes of the Rockies." The outflow of Granby Lake is considered the Headwaters. This high mountain country, over 8,000 feet in elevation, offers excellent fishing, even for Greyling in Shadow Mountain Lake. Varied water sports and other outdoor activities from backpacking to glider soaring are available. Grand Lake is the home of the world's highest elevation Yacht Club. There is an abundance of campgrounds in the National Recreation Area and the Arapaho and Roosevelt National Forests as well as at private campgrounds and resorts throughout the area.

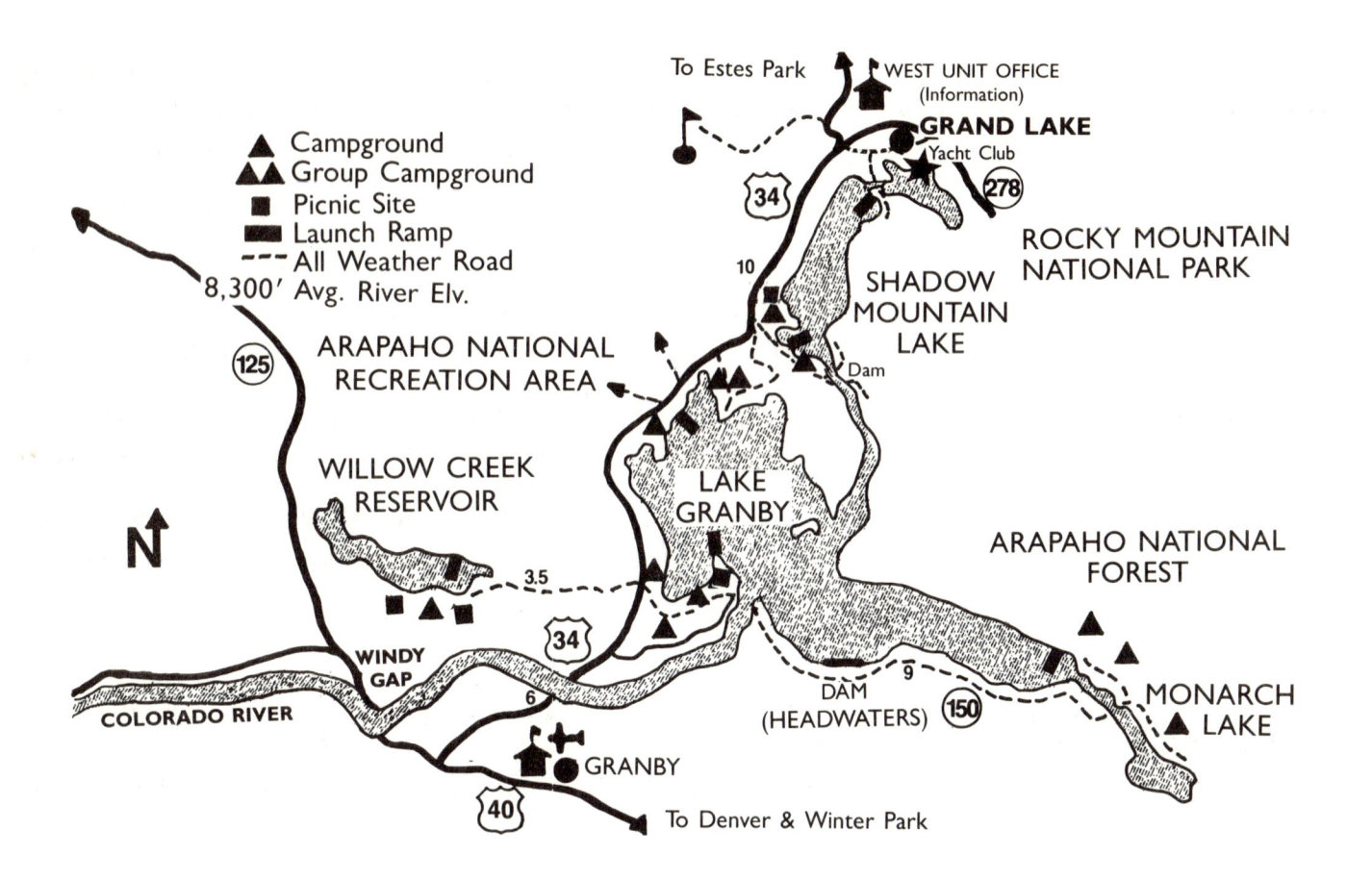

INFORMATION: Sulphur Ranger District, Star Rt., Granby, CO 80446, Ph: 303-877-3331		
CAMPGROUNDS & RESORTS	**RECREATION**	**OTHER**
See Following Page	Boating: Canoeing and All Types of Boating on the Lakes (Boats with Motors Over 10HP Must Have State License), Rafting, Wind Surfing. Fishing: Rainbow, Cutthroat, Kamloop, Brook, Brown and Mackinaw Trout, Silver and Kokanee Salmon, Greyling. Hunting: Elk, Deer and Bear Hiking and Riding Trails Picnic Areas and Nature Study	Marinas Golf and Tennis Winter Sports—Snowmobiling and Ice Fishing Complete facilities in Grand Lake and Granby

ARAPAHO NATIONAL RECREATION AREA

Arapaho Bay Campground—77 Sites, vault toilets, drinking water, tables, firegrates and launch ramp. Fee: $5.

Green Ridge Campground—83 Sites, modern comfort stations, running water, tables, firegrates, launch ramp and disposal station. Fee: $7.

Stillwater Campground—145 Sites, modern comfort stations, running water, tables, firegrates, amphitheater, launch ramp and disposal station. Fee: $7.

Willow Creek Campground—35 Sites, vault toilets, drinking water, firegrates, tables and launching area for small boats. Fee: $5.

For additional public campgrounds, dispersed camping areas or information, contact the Sulphur Ranger District.

Winding River R.V. Campground—P.O. Box 629, Grand Lake 80447, Ph: 303-627-3215.

This 160 acre private campground has 165 sites for tents and R.V.s, some with electric and sewer hookups. The family campground offers horseback riding, hayrides, ice cream socials and Disney movies. Open from May 30 to October 15.

Gala Marina—North Shore Lake Granby, Ph: 303-627-3220.

Boat rentals, Sales and Service, Housekeeping Units.

Norton Marina—Hwy. 34 and Lake Granby, Ph: 303-887-3456.

Full Service Marina, Boat Tours and Service, Fishing and Boating Supplies, Snack Bar and Overnight Camping.

Trail Ridge Marina—Shadow Mountain Lake, 2½ Miles South of Grand Lake

Boat Rentals and Service, Fishing and Boating Supplies, Snack Bar and Overnight Camping.

Lake Kove Marina—P.O. Box 469, Shadow Mountain Lake, Ph: 303-627-3605.

Boat Rentals, Sales and Service, Snack Bar, Fishing and Boating Supplies.

Grand Lake Marina—P.O. Box 530, Grand Lake, Ph: 303-627-3401.

Home of the Classic Antique Cris Crafts, Tours of Grand Lake, Dinner Cruises, Private Charters, Complete Line of Rentals, Sales, Service, Storage and Docking.

For Additional Information Concerning Resorts and Facilities in This Area, Contact:
Grand Lake Chamber of Commerce
Box 57
Grand Lake, CO 80447
Ph: 303-627-3402

GRANBY TO TROUBLESOME INCLUDING
HOT SULPHUR SPRINGS

The Colorado River begins its whitewater journey through this scenic Rocky Mountain backcountry. The Colorado Wildlife Commission has designated the River from Windy Gap twenty miles downstream to Troublesome Creek, a "Gold Medal Stream." The fishing is outstanding for large trout. Williams Fork Reservoir offers the angler Northern Pike. Granby, the western gateway to the Arapaho National Recreation Area, is the hub of numerous recreational facilities and opportunities. Although campgrounds are limited to the Recreation Area and National Park, there are numerous privately operated facilities in the surrounding areas. Hot Sulphur Springs offers one of the finest mineral baths in the nation. Across the River, the historic Riverside Hotel provides an excellent restaurant and pleasant rooms with some of them overlooking the Colorado River.

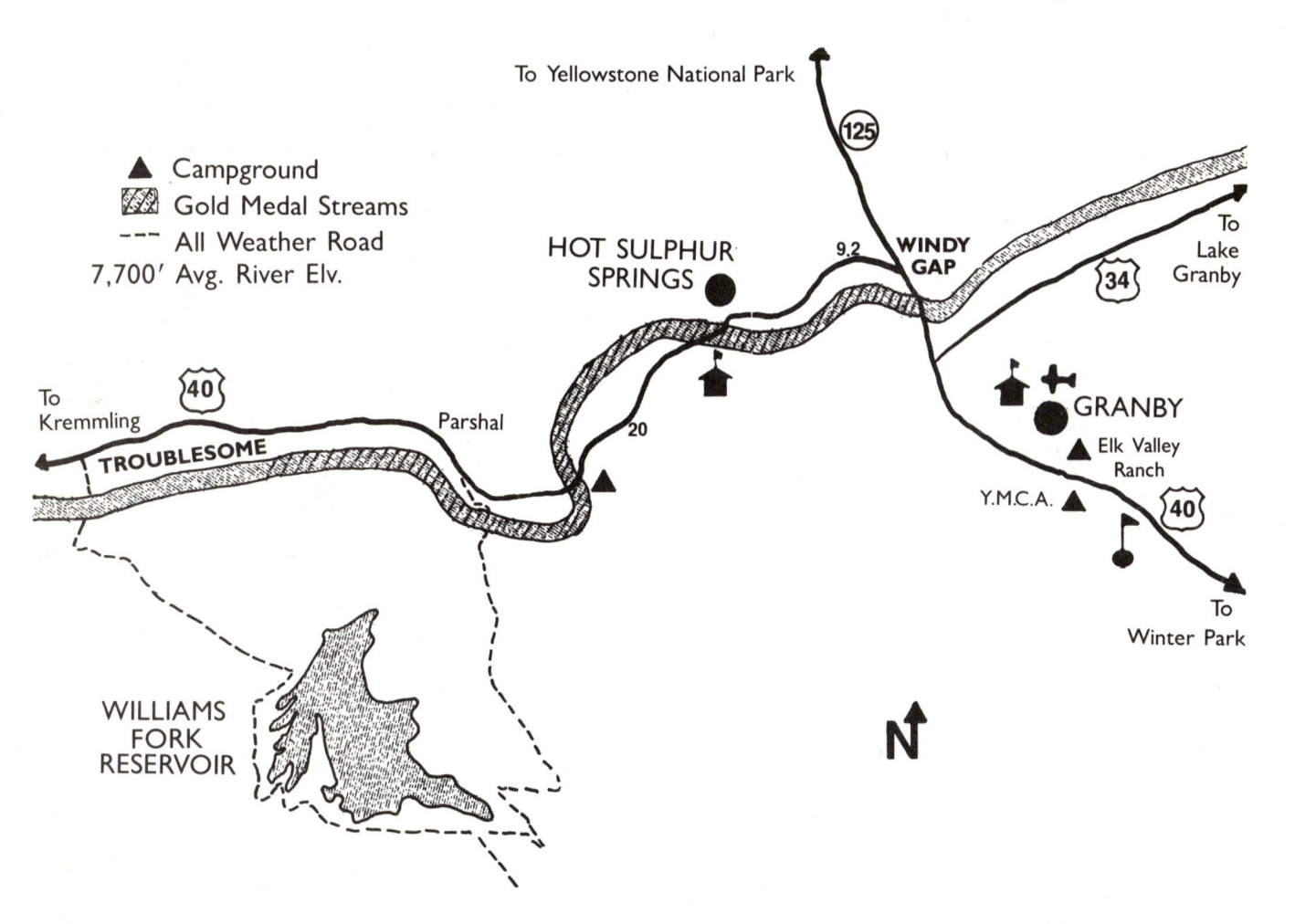

INFORMATION: Granby Chamber of Commerce, 200 E. Agate Ave., Granby, CO 80446, Ph: 303-887-2311

CAMPGROUNDS & RESORTS

YMCA of the Rockies—Snow Mountain Ranch, P.O. Box 558, Granby, CO 80446, Ph: 303-443-4743. 62 Tent & R.V. Sites, Some Hookups, 41 Cabins, Indoor Pool

Elk Valley R.V. Campground—P.O. Box 1142, Granby, CO 80446. 9 Sites, Full Hookups

Contact Chamber of Commerce for full info.

RECREATION

Rafting
Fishing: Rainbow & Brown Trout—"Gold Medal Water," Northern Pike
Hunting: Deer, Bear & Elk
Hiking & Riding Trails
Backpacking & Pack Trips
Rock Hounding
4-Wheel Trails
Nature Walks
Golf & Tennis
Dude Ranches

OTHER

Silver Creek Resort—P.O. Box 636, Granby, CO 80466, Ph: 303-887-2755 or for Reservations: Ph: 303-887-2189.

Hot Sulphur Mineral Baths—5609 Grand County Rd. 20, Hot Sulphur Springs, CO 80451, Ph: 303-725-3306.

Riverside Hotel—Hot Sulphur Springs, CO 80451, Ph: 303-725-9996 or 303-725-3589.

TROUBLESOME TO STATE BRIDGE
INCLUDING KREMMLING

Paralleled by graveled country roads and the Denver and Rio Grande Railroad, this section of the Upper Colorado River wanders amid its natural setting of agricultural lands and mesas. Relatively mild rapids (Class I and II), few hazards, easy access and numerous campgrounds make this a popular rafting area. Big Gore Canyon is not advised for float boating. Only the advanced whitewater canoer should attempt the rapids between Pumphouse and Radium. The novice and intermediate can enjoy the waters below Radium. The River from Big Gore Canyon to State Bridge has been designated by the Colorado Wildlife Commission as "Wild Trout Waters," so the angler will find the challenge and rewards of native wild trout. The majority of the facilities are under the jurisdiction of the Bureau of Land Management. The Forest Service operates three campgrounds at Green Mountain Reservoir, 10 miles south of Kremmling on SR 9.

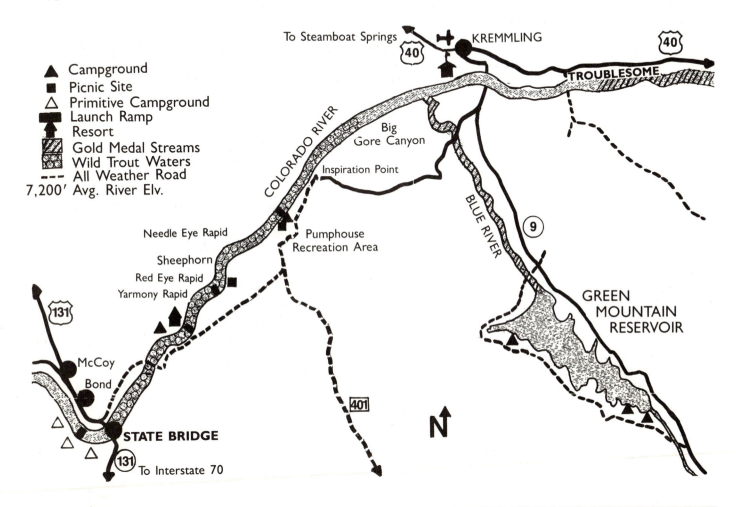

INFORMATION: Bureau of Land Management, P.O. Box 68, Kremmling, CO 80459, Ph: 303-724-3437

CAMPGROUNDS & RESORTS	RECREATION	OTHER
BLM Campgrounds as Noted on Map **U.S. Forest Service**—Dillon Ranger District, P.O. Box 498, Blue River Center, Silverthorne, CO 80489, Ph: 303-468-5400. Elliott Creek: 64 Sites McDonald Flats: 14 Sites Prairie Point: 39 Sites	Boating: Green Mountain Reservoir Rafting: Canoeing, Kayaking Fishing: Rainbow, Brown, Cutthroat and Brook Trout, Mountain White Fish, Wild Trout Waters as shown on map. Hunting: Bear, Deer and Elk. Hiking Trails Picnicking and Swimming Bird Watching & Nature Study 4-Wheel Trails.	Facilities in Kremmling **Rancho Del Rio**—Resort Grocery Store & Phone **Bond**—Restaurant, Gas Station & Phone Post Office

STATE BRIDGE TO DOTSERO TO
BORDER OF GLENWOOD CANYON

This section of the Upper Colorado is similar to its neighbor up river in recreational opportunities and natural terrain. The Upper Colorado remains relatively unchanged since its original settlement in the late 1800's with its log cabins and the waterwheel near McCoy. While tubing is not advised, canoers will find this area an ideal place to develop whitewater skills. Canoers are advised to scout and portage in waters beyond their skills. There are numerous campsites and access points along the River for one day trips or more. River access beyond Dotsero is limited. This section is less heavily used than the more congested area above State Bridge.

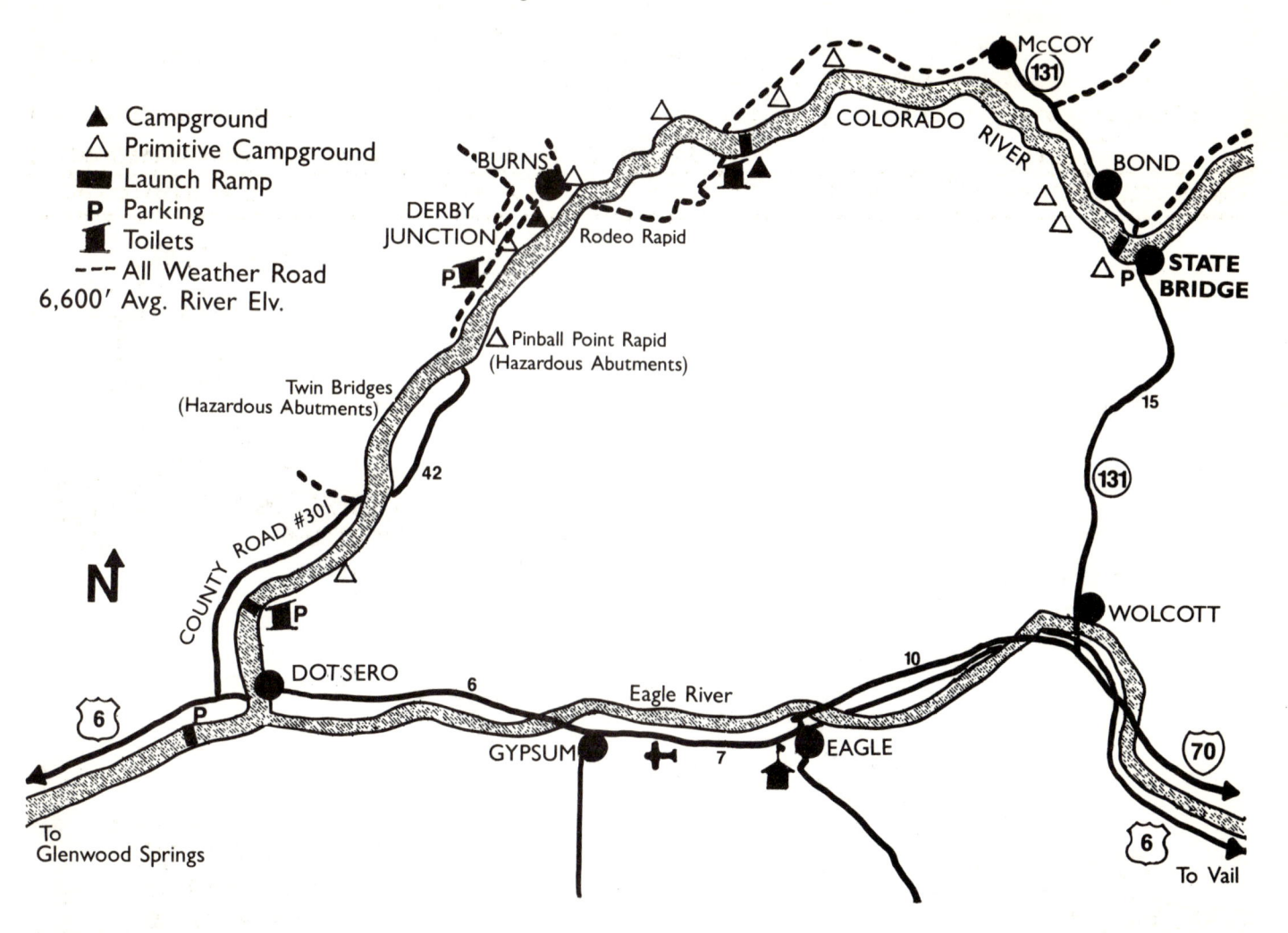

INFORMATION: Bureau of Land Management, P.O. Box 1009, Glenwood Springs, CO 81601, Ph: 303-945-2341

CAMPGROUNDS & RESORTS	RECREATION	OTHER
BLM Campgrounds as Noted on Map.	Rafting, Canoeing and Kayaking Fishing: Rainbow and Brown Trout Picnicking Swimming Nature Study Bird Watching Hiking Trails 4-Wheel Trails	**Bond** Restaurant, Gas, Post Office. **Burns** Grocery Store, Gas, Post Office. **Derby Junction** Store, Raft Rentals, Cabins and Camping

GLENWOOD CANYON AND GLENWOOD SPRINGS

The beauty and power of the mighty Colorado are proudly displayed as it carves its way down the eighteen miles of Glenwood Canyon through the White Mountain National Forest. Sheer walls of varied hues rise up a thousand feet above this breathtaking corridor through the Rockies. Just west of the Canyon, historic Glenwood Springs provides an urban setting within this area of natural abundance. Native Americans, centuries before the white man arrived, migrated to this sacred spa at the confluence of the Yampah (Big Medicine) Hot Springs. Glenwood Springs is still popular with its world famous open air swimming pool and therapeutic vapor caves. You can visit "Doc" Holliday's memorial at Cemetery Hill or visit "Teddy" Roosevelt's 1905 Spring White House at the Hotel Colorado. Be sure to ask about the legend of the "Teddy Bear."

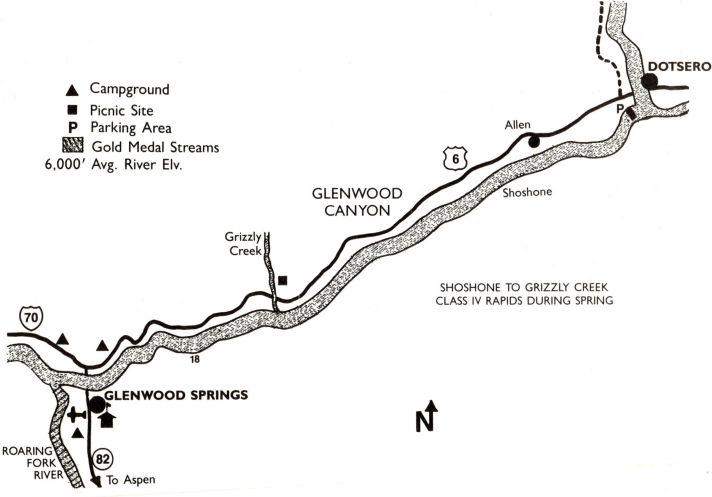

▲ Campground
■ Picnic Site
P Parking Area
▨ Gold Medal Streams
6,000' Avg. River Elev.

DOTSERO

Allen

(6)

Shoshone

GLENWOOD CANYON

Grizzly Creek

SHOSHONE TO GRIZZLY CREEK
CLASS IV RAPIDS DURING SPRING

(70)

18

N

GLENWOOD SPRINGS

ROARING FORK RIVER

(82)

To Aspen

INFORMATION: Glenwood Springs Chamber of Commerce, 1102 Grand Ave., Glenwood Springs, CO 81601, Ph: 303-945-6589		
CAMPGROUNDS & RESORTS	**RECREATION**	**OTHER**
Numerous Campgrounds and Resorts See Following Page.	Fishing: Rainbow, Brown, Cutthroat, Brook and Mountain White Fish. Hunting: Elk, Mule Deer, Black Bear, Grouse, Waterfowl. Rafting, Canoeing, Kayaking Picnicking & Backpacking Hiking & Climbing Horseback Riding Swimming, Golf & Tennis Jeep Tours & 4-Wheel Trails Winter Sports	Full Facilities in Glenwood Springs Contact Chamber of Commerce

GLENWOOD CANYON AND GLENWOOD SPRINGS

Campgrounds

Ami's Acres Campground—P.O. Box 1239, W. Glenwood Springs 81602, Ph: 303-945-5340.
66 R.V. Sites, Full Hookups, Tent Area, Motorcyclists Invited, Showers, Laundry.

The Hideout Campground—1293 CR 117, Glenwood Springs 81601, Ph: 303-945-5621.
60 Sites, 24 Full Hookups, Tent Area, Cabins, Motorcyclists Invited, Disposal Station, Showers.

Rock Gardens Campground—1308 CR 129, Glenwood Springs 81601, Ph: 303-945-6737.
36 R.V. Sites, 39 Tent Sites, Partial Hookups, Showers, Store, Raft Trips—Reservations Advised.

Resorts

Hotel Colorado—526 Pine Street, Glenwood Springs 81602, Ph: 303-945-6511.
Listed in the National Registry of Historic Places, 114 Rooms and Suites, Restaurant, Bar, Conference Rooms
Glenwood Activities Center: Blue Sky Adventures, Ph: 303-945-6605.
Health Spa: Fitness Center, Ph: 303-945-8107.

Glenwood Hot Springs Lodge—401 N. River Street, P.O. Box 308, Glenwood Springs 81601, Ph: 303-945-6571.
Rooms and Family Suites, Restaurant, Bar, World's Largest Hot Springs Pool, Waterslide, Athletic Club, Sport Shop, Miniature Golf.

Glenwood Springs Vapor Caves—709 E. 6th Street, Glenwood Springs 81602, Ph: 303-945-5825.
Caves offer Natural Steam Baths at 115° with 100% Humidity. Therapeutic Massages, Facials and Reflexology.

GLENWOOD SPRINGS TO PALISADE

Continuing its westward journey in this area of mineral abundance, the Colorado River flows past the cascading rapids of Rifle and Plateau Creeks down through De Beque Canyon and into the rich orchards of the Grand Valley at Palisade. Recreational opportunities abound. Rifle Gap-Falls State Recreation Area provides a scenic diversion with its three falls, natural caves and man-made reservoir (warm water fishery). Grand Mesa, the world's largest flattop mountain, 40 miles long, offers lake fishing, streams and trails, support facilities and the Powderhorn Ski Area. Next to the Grand Mesa National Forest, Vega State Recreation Area provides a 900 acre Lake for fishing, swimming, boating and camping.

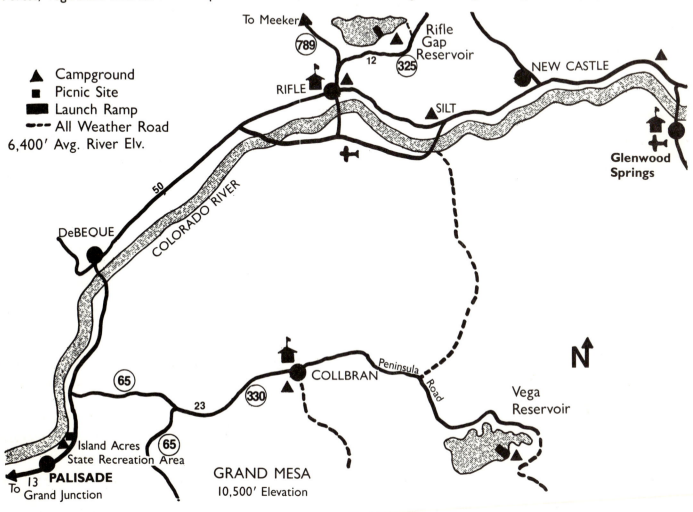

INFORMATION: Colorado DPOR—W. Regional Office, State Service Bldg., Grand Junction, Co 81501, Ph: 303-248-7319

CAMPGROUNDS & RESORTS	RECREATION	OTHER
Rifle Gap-Falls State Rec. Area 70 Sites, Disposal Station, Launch Ramp, Ph: 303-625-1607 **Island Acres State Rec. Area** 32 Sites, Disposal Station, Ph: 303-464-0548 **Vega State Recreation Area** 110 Sites, Launch Ramp, Ph: 303-487-3407 **Sievers Wild Goose Camp Park** 78 Sites, R.V. Hookups, Ph: 308-876-2443 **Elk Creek Campground** 30 Sites, Ph: 303-984-2257	Rafting, Canoeing, Kayaking Boating: Rifle Gap & Vega Lakes Fishing: Rainbow, Brown, Cut-throat & Brook Trout, Channel Catfish, Walleye, Large & Smallmouth Bass Backpacking, Hiking & Riding Trails, Rock Climbing Hunting: Bear, Deer & Elk Swimming Nature Study Picnic Areas Winter Sports	**Rifle Ranger District** 1400 Access Road, Rifle, CO 81650, Ph: 303-963-2266 **Collbran Ranger District** 212 E. High Street, Collbran, CO 81624 **Grand Junction Ranger District** Federal Bldg., Grand Junction, CO 81501

GRAND JUNCTION

Grand Junction is the County seat of Mesa County which has the highest population in Western Colorado. Located at the junction of Colorado's two greatest rivers, the Colorado and the Gunnison, this urban area serves as the industrial, cultural and transportation hub for Western Colorado and Eastern Utah. The Grand Valley, once an inland sea and tramping ground for dinosaurs, Indians and prospectors, is surrounded by the Grand Mesa, the Bookcliffs and the Colorado National Monument, providing a colorful living geological history. It is the gateway to an abundance of scenic and recreational opportunities from a picnic on the rim of the National Monument, camping on the Grand Mesa or fishing in one of the 200 sparkling lakes or streams nearby.

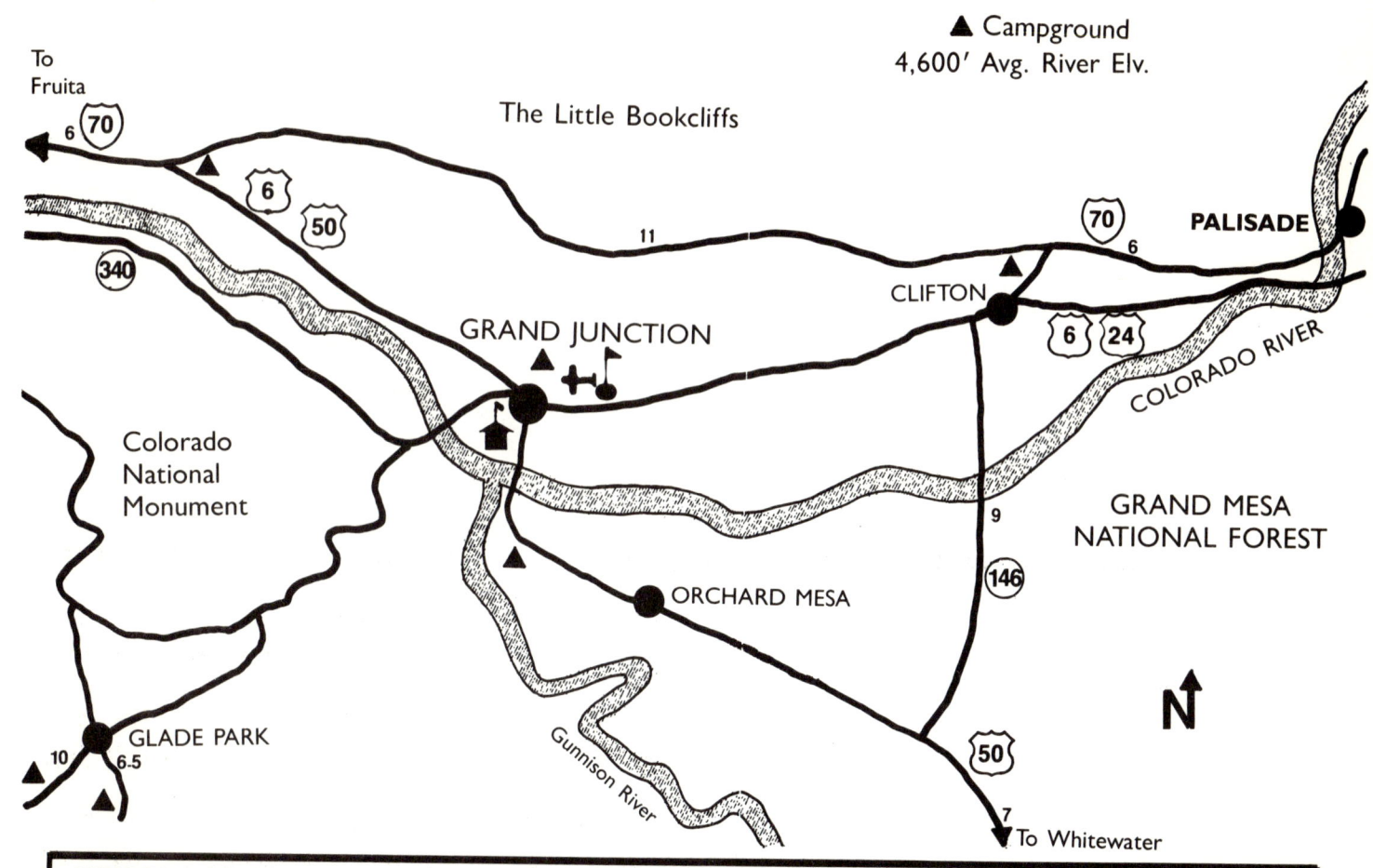

▲ Campground
4,600' Avg. River Elv.

INFORMATION: Grand Junction Area Chamber of Commerce, 360 Grand Ave., P.O. Box 1330, Grand Junction, CO 81502, Ph: 303-242-3214

CAMPGROUNDS & RESORTS	RECREATION	OTHER

Numerous Campgrounds
 	
See Following Page
	Rafting, Tours and Boating
	Fishing:
	   Rivers: Channel Catfish
	   Lakes and Streams: Rainbow,
	     Cutthroat and Lake Trout
	Hunting: Elk, Deer, Bear, Pheasant
	   and Waterfowl
	Picnicking
	Swimming
	Hiking and Nature Trails
	Backpacking & 4-Wheel Touring
	Horseback Riding
	Tennis & Golf
		Full Facilities in Grand Junction
		
		Contact Chamber of Commerce Ph:
		   303-242-3214

CAMPGROUNDS AND RESORTS

Big J Camper Court—2819 Highway 50 East, Grand Junction, Co 81501, Ph: 303-242-2527.

2 Miles South of Grand Junction, 160 R.V. Sites, Full Hookups, Tents, Propane, Laundry, Showers, Pool, Recreation Room, Playground.

Grand Junction/Clifton KOA Campground—3238 F Road, Clifton, CO 81520, Ph: 303-434-6644.

On Highway I-70 at Exit 37, 141 R.V. Sites, Full Hookups, Laundry, Showers, Pool, Recreation Room, Playground, Groceries, **No** Pets.

Mobile City Mobile Park—2322 Highway 6 and 50, Grand Junction, Co 81501, Ph: 303-242-9291.

3 Miles West of Grand Junction, 40 R.V. Sites, Full Hookups, Laundry.

Rose Park Mobile Village—2910 North Avenue,#22A, Grand Junction, CO 81501, Ph: 303-243-1292.

On East Site of Town, 26 R.V. Sites, Full Hookups, Laundry, Showers.

Mud Spring Campground—Bureau of Land Management.

6½ Miles South of Glade Park Store on CR 16.5, 10 Sites, No Hookkups.

Little Dolores Falls—Bureau of Land Management.

10 Miles Southwest of Glade Park Store, 19 Sites, No Hookups.

Powderhorn Ski Area—On the Grand Mesa off Colorado 65. Lodging, Campsites nearby. No Hookups.

SOUTHWEST OF GRAND JUNCTION TO THE UTAH BORDER

Leaving Grand Junction, the Colorado River flows past one of the world's natural wonders on its westward journey to Utah. The Colorado National Monument with its 18,000 acre natural amphitheater provides a spectacular view of millions of years of geological history. Wandering through this area of desert canyons, the River provides year round opportunities for canoeing, kayaking and rafting. Float boating is not advised above the Loma launch site. From the Loma Boat Launch to the Westwater Ranger Station, this 26 mile section of the River offers a leisurely scenic float through Horsethief and Ruby Canyons and is the starting point for many professionally guided, action whitewater trips. The boater should be aware that travel time can vary (6 to 15 hours) due to wind and water conditions, so be prepared to spend the night.

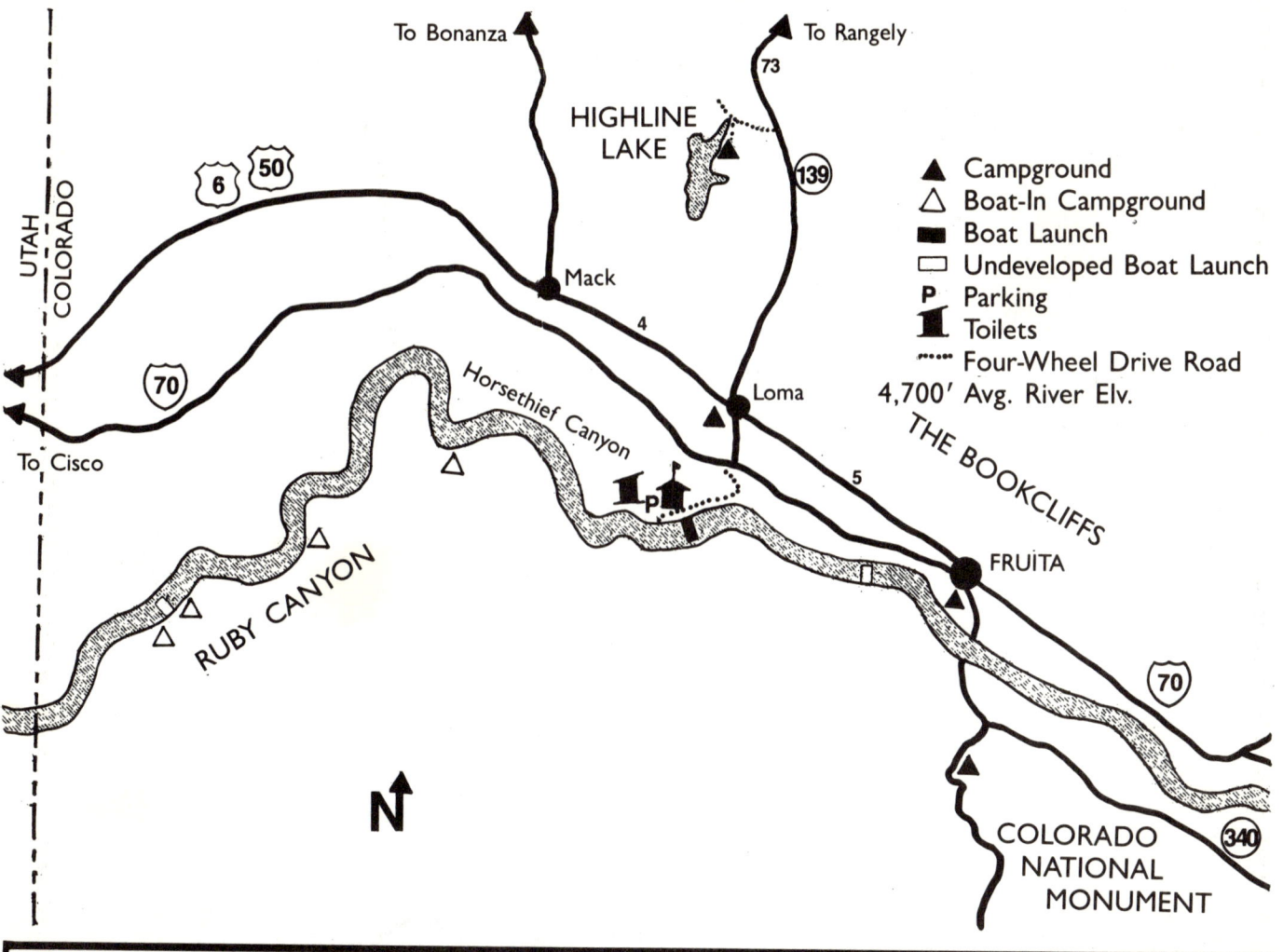

| INFORMATION: Grand Junction Area Chamber of Commerce, 360 Grand Ave., P.O. Box 1330, Grand Junction, CO 81502, Ph: 303-242-3214 |

CAMPGROUNDS & RESORTS	RECREATION	OTHER
Fruita Junction R.V. Park—607 Highway 340, Fruita, CO, Ph: 303-858-3155 **Highline State Recreation Area** 25 Tent/R.V. Sites, No Hookups, Disposal Station, Picnic Units, Swim Beach	Boating: Highline Lake—Launch Ramp & Waterskiing Float Boating—No Rapids Fishing: River—Channel Cats; Lakes & Streams—Rainbow Hunting: Elk, Bear, Deer and Waterfowl Backpacking & Hiking Nature Study & 4-Wheel Tours	**Colorado National Monument** National Park Service Superintendent Fruita, CO 81521 Campground: Saddlehorn: 81 Tent/R.V. Sites, Picnic Areas, Nature Trails, Visitor's Center, Ph: 303-858-3617 Full Facilities in Grand Junction

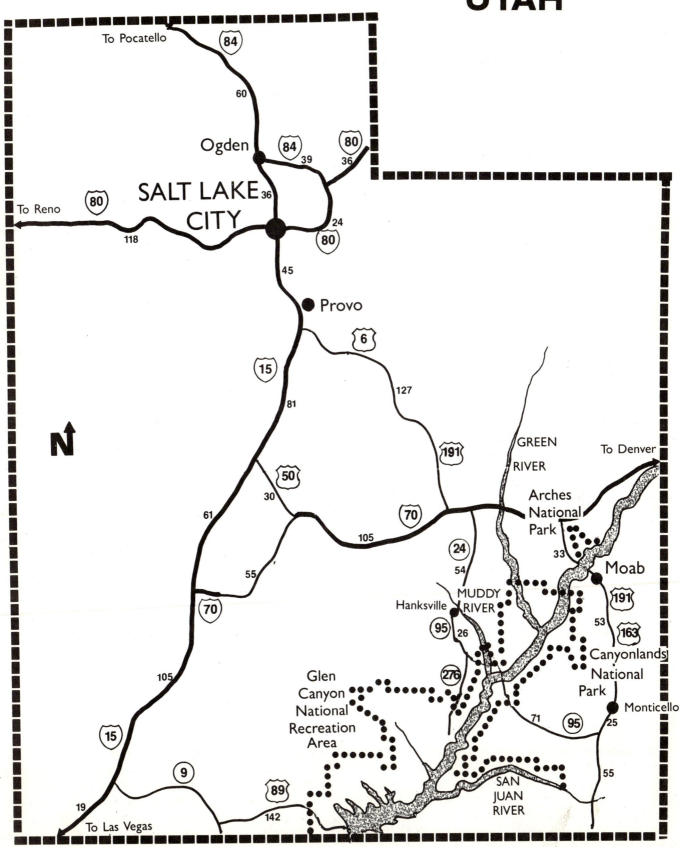

UTAH

To Pocatello
84
60
Ogden
84 39
80 36
SALT LAKE 36
CITY
To Reno
80
118
80 24
45
Provo
6
15
81
127
191
GREEN
RIVER
To Denver
50
Arches
National
Park
30
70
33
105
Moab
24
191
54
53
MUDDY
RIVER
163
Hanksville
Canyonlands
95 26
National
Park
70
276
Monticello
Glen
Canyon
National
Recreation
Area
71
95 25
105
15
55
9
SAN
JUAN
RIVER
89
15
19
142
To Las Vegas

N

NO TRACE CAMPING

No Trace Camping is the art of using the land and leaving it exactly as you found it. Our outdoor environments and experiences cannot survive without it. While standards vary and depend on the characteristics of the land, vegetation, climate and management, there are some general rules to follow.

Human waste can cause serious problems, such as Giardiasis, a painful intestinal disorder. While toilets are provided in some areas along the River, there are many stretches where none are available. Many sections require a portable toilet and you are required to deposit the waste at designated recepticals. Where groups are small and use is slight, a "cat hole" is the appropriate method. Dig a five- to eight-inch hole at least 100 feet from water. Dig a latrene for larger groups. Fill these holes before you leave and burn all paper products.

Personal hygiene and washing with soap and water can prevent serious disease problems especially before handling food. Always use biodegradable soaps away from the River. Bathe and wash dishes at least 100 feet from the River, and deposit the dirty soapy water into the "cat hole" or latrine.

Campfires too often result in an eyesore with blackened rocks and partially burned firewood. Charcoal covered beaches too often are scarred with aluminum foil and other non-combustibles. The best way to prevent these eyesores is to use a portable stove, but if you must build a fire and it is permitted, use a fire ring. The ashes may be carried out. Use only dead wood or driftwood, if allowed. Build the fire in a safe place and never leave it unattended. Keep non-burnables out of the fire, and scatter unused firewood.

All trash should be carried out. If you have a campfire, burn all combustibles. Never throw cans or bottles into the River. If you smoke, put the butts into a trash bag. Do not bury garbage as animals can dig it up. Make a final check before leaving camp and be sure to take all waste with you.

Regulations vary for each section of the River. Always check with the local Ranger for specification information.

COLORADO-UTAH BORDER THROUGH WESTWATER CANYON

Leaving Ruby Canyon, the Colorado takes a leisurely southwestern turn into the Canyonlands section of Utah. This primitive area of geological splendor offers few facilities beyond nature's rugged abundance. Those that are provided by the Bureau of Land Management are related to float boating, and these are even sparse. There are opportunities for the angler, hunter and 4-wheeler, but this is primarily rafting country. While canoers enjoy the River above and below Westwater, this world-famed Canyon is only for the **expert** kayaker and rafter. Permits and reservations are required between May and September on this popular section of the River.

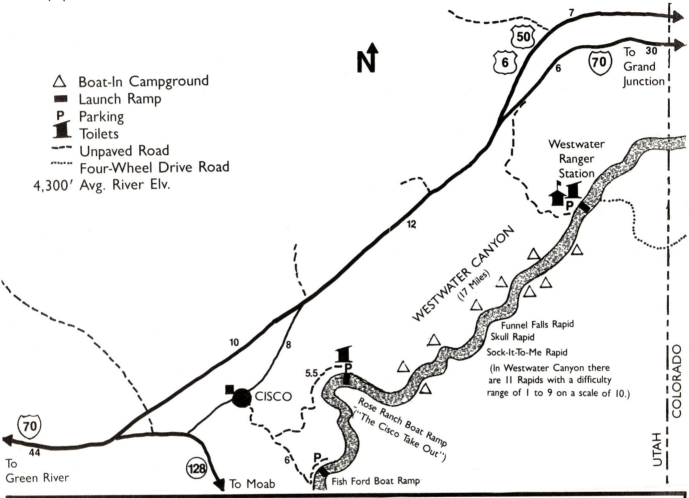

△ Boat-In Campground
■ Launch Ramp
P Parking
🚻 Toilets
- - - Unpaved Road
······ Four-Wheel Drive Road
4,300′ Avg. River Elv.

N

To Grand Junction

Westwater Ranger Station

WESTWATER CANYON (17 Miles)

Funnel Falls Rapid
Skull Rapid
Sock-It-To-Me Rapid

(In Westwater Canyon there are 11 Rapids with a difficulty range of 1 to 9 on a scale of 10.)

Rose Ranch Boat Ramp ("The Cisco Take Out")

CISCO

Fish Ford Boat Ramp

To Green River

To Moab

UTAH   COLORADO

INFORMATION: Bureau of Land Management - Grand Area Office, P.O. Box M, Sand Flat Rd., Moab, UT 84532
Ph: 801-259-8193

CAMPGROUNDS & RESORTS	RECREATION	OTHER

Westwater Canyon: Permit Required.
 	
Camping Allowed on Public Lands
 	
Boat Camping in Westwater Canyon Subject to Permit and Regulations—Contact Bureau of Land Management in Moab
	Float Boating:
	  Westwater Canyon: Class IV
	  Expert—Reservations & Permits Required—2 Months in Advance—Contact BLM
	Fishing: Channel Catfish, Black Bullhead, Largemouth Bass and Bluegill
	Hunting: Deer and Waterfowl
	Hiking & 4-Wheeling
	Nature Study
	Artifacts: Look But Don't Take
		Major Facilities and Lodging in Grand Junction and Moab
		
		Commercial Float Boating and Tours

SOUTHWEST OF WESTWATER CANYON TO ARCHES NATIONAL PARK

Leaving the whitewater excitement of Westwater's granite canyon, the Colorado opens into a broad valley presenting a distant view of the snow capped La Sals. Meandering through this area of hay fields and grazing lands, the River absorbs the flow of the Dolores River. The Dolores usually offers the float boater a swift, free flowing River in Spring and Summer. Below historic Dewey Bridge, the Colorado provides the boater and motorist with a popular scenic experience. Utah Highway 128 parallels the River past the settings of numerous motion pictures, side canyons and the dramatic series of redrock spires of Fisher Towers.

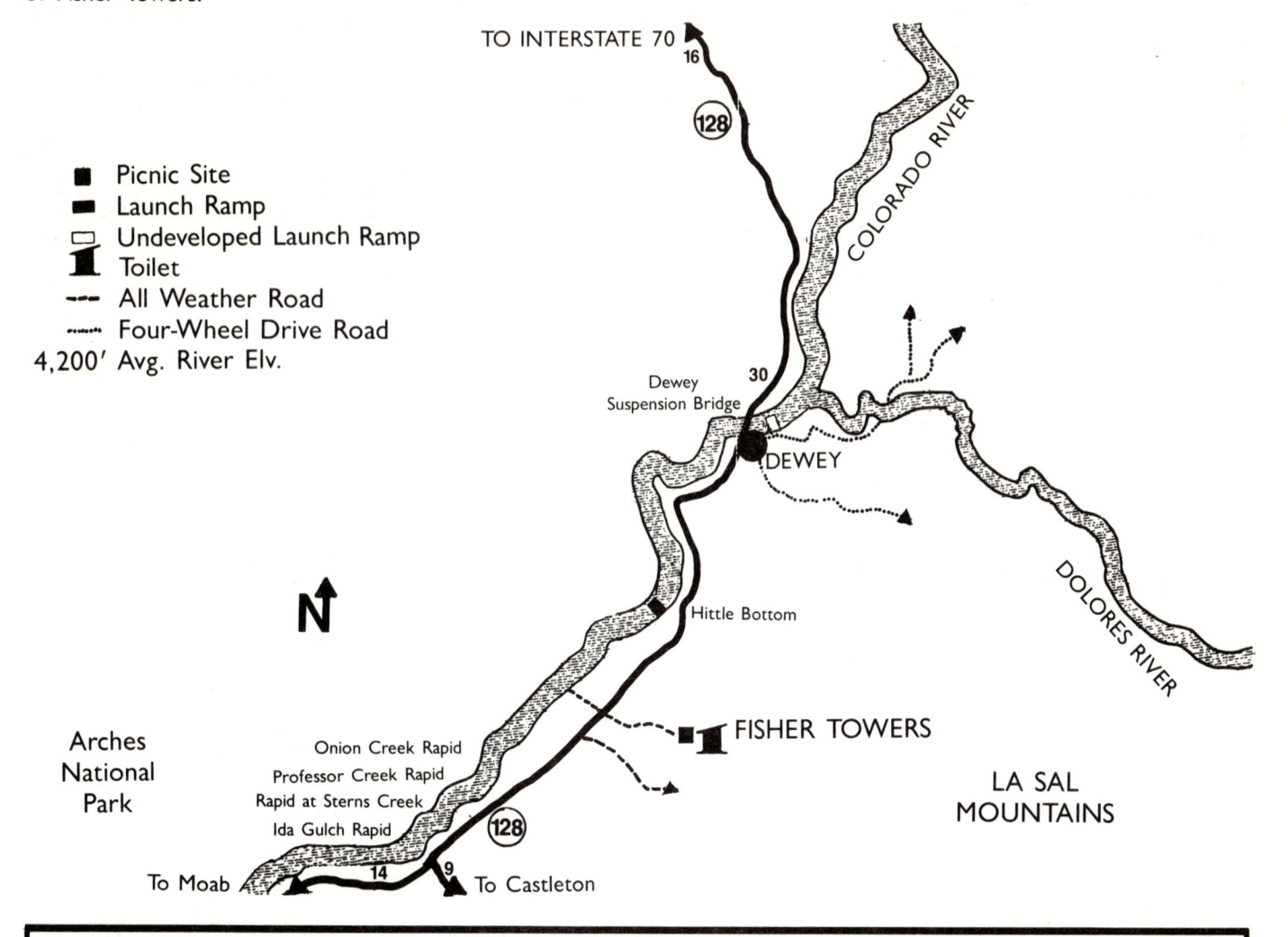

INFORMATION: Bureau of Land Management - Grand Area Office, P.O. Box M, Sand Flat Rd., Moab, UT 84532
Ph: 801-259-8193

CAMPGROUNDS & RESORTS	RECREATION	OTHER

Limited Primitive Sites on Public Lands along River
Fire Pans and Portable Toilets Required
	Float Boating
	  Colorado River: Class II White Water
	  Dolores River: Permits Required— Class IV; Mid May to Mid June
	  River Flow Information:
	    Ph: 801-524-5130
	Fishing: Channel Catfish, Black Bullhead, Largemouth Bass, Bluegill
	Hunting: Deer and Waterfowl
	Hiking & 4-Wheeling
		Major Facilities in Moab

ARCHES NATIONAL PARK AND MOAB

Moab is the only Utah City on the Colorado River. Nestled amid red desert cliffs and the distant snow-capped La Sal Mountains, Moab presents a beautiful scene of nature's contrasts. This area was once the headquarters of the famed Butch Cassidy's outlaw gang and the site of many Zane Grey novels. Often called the "Heart of the Canyonlands," Moab is the hub of numerous scenic and recreational opportunities. While whitewater boating tours are a major attraction, there is an abundance of other opportunities from a horseback trip into Arches National Park to a scenic flight over the beautiful Canyonlands. There is even the scenic Slickrock Bike Trail for trail motor bikes. This is administered by the Bureau of Land Management. Arches National Park features the world's greatest concentration of natural stone arches. The view from Windows Section of the Park is spectacular.

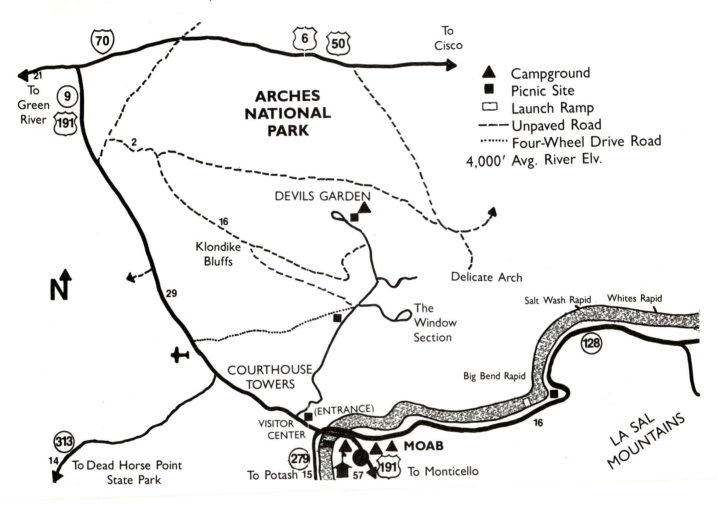

INFORMATION: Grand County Travel Council, 805 N. Main, Moab, UT 84532, Ph: 801-259-8825

CAMPGROUNDS & RESORTS	RECREATION	OTHER
Arches National Park Superintendent, Canyonlands National Park, 125 W. 200 South Moab, UT 84532, Ph: 801-259-8161. **Devil's Garden Campground** 54 Tent/R.V. Sites, Water May-October Only **KOA** 3225 So. Hwy. 191, Moab, UT 84532, Ph: 801-259-6682, 54 Sites, Full Hookups, Disposal Station, Laundry, Store, Pool	Rafting: Calm & Whitewater Fishing: Channel Catfish, Black Bullhead Hiking & Riding Trails Picnicking Archeological & Geological Study Tours: Van, Jeep, Float & Jet Boat, Plane, Horseback Moab—Slickrock Bike Trail	**Canyonland Campark** 555 S. Main, Moab, UT 84532, Ph: 801-259-6848, 113 Tent/R.V. Sites, Full Hookups, Disposal Station, Pool, Laundry, Store, Restaurant **Slickrock Country Campground** 1301½ N. Highway 191, Moab, UT 84532, Ph: 801-259-7660, 144 Tent/R.V. Sites, Full Hookups, Disposal Station, Pool, Laundry, Store, Restaurant

ARCHES NATIONAL PARK
By Lin Ottinger

Arches National Park—73,379 acres, bordering the Colorado River on the south end of several miles. The park has the world's largest concentration of natural stone arches. These arches have been formed through millions of years of erosion, the wind and water slowly dissolving the soft, porous sandstone. One of these arches spans a distance of 291 feet, the longest known natural stone arch. Many of these oddities of nature are hidden in the miles of remote canyons which abound in the park. The canyons contain arches rarely seen by anyone, and no doubt some are still left undiscovered.

Although there are good paved roads throughout the park, which enables the visitor to view more Arches on a 20-mile drive than anywhere else, there remains hidden amongst the numerous balanced rocks and sandstone spires, many beautiful Arches of all sizes and shapes, many only a few hundred feet from the road. There are miles of foot trails well marked and maintained by the park service. Some of these trails are only a few hundred feet long while others are many miles. There remains in this vast expanse many areas virtually unexplored.

This map was published by Lin Ottinger, artwork by Jean-Claude Gal and is available printed on poster quality paper 18 x 22 inches at:

Lin Ottinger Tours
Moab Rock Shop
137 N. Main Stret
Moab, Utah 84532

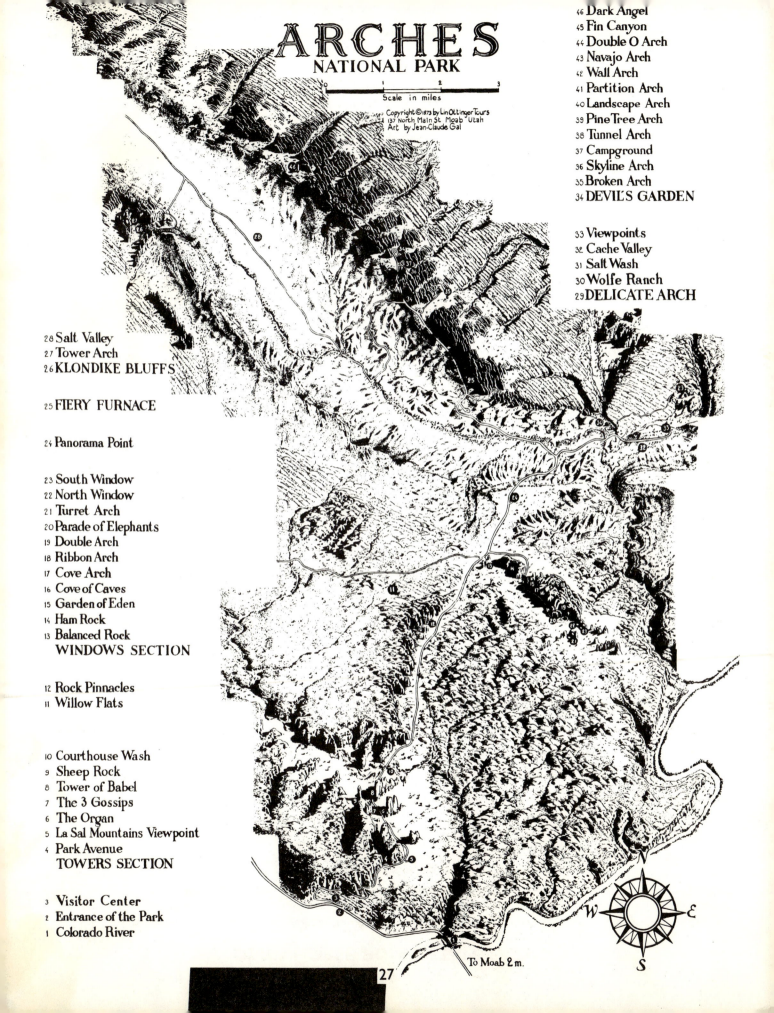

ARCHES
NATIONAL PARK

Scale in miles

Copyright © 1973 by Lin Ottinger Tours
137 North Main St. Moab Utah
Art by Jean-Claude Gal

46 Dark Angel
45 Fin Canyon
44 Double O Arch
43 Navajo Arch
42 Wall Arch
41 Partition Arch
40 Landscape Arch
39 Pine Tree Arch
38 Tunnel Arch
37 Campground
36 Skyline Arch
35 Broken Arch
34 DEVIL'S GARDEN

33 Viewpoints
32 Cache Valley
31 Salt Wash
30 Wolfe Ranch
29 DELICATE ARCH

28 Salt Valley
27 Tower Arch
26 KLONDIKE BLUFFS

25 FIERY FURNACE

24 Panorama Point

23 South Window
22 North Window
21 Turret Arch
20 Parade of Elephants
19 Double Arch
18 Ribbon Arch
17 Cove Arch
16 Cove of Caves
15 Garden of Eden
14 Ham Rock
13 Balanced Rock
WINDOWS SECTION

12 Rock Pinnacles
11 Willow Flats

10 Courthouse Wash
9 Sheep Rock
8 Tower of Babel
7 The 3 Gossips
6 The Organ
5 La Sal Mountains Viewpoint
4 Park Avenue
TOWERS SECTION

3 Visitor Center
2 Entrance of the Park
1 Colorado River

27

To Moab 2 m.

DEAD HORSE POINT STATE PARK

Still relaxing from its whitewater fury at Westwater Canyon, the Colorado River provides for a quiet excursion in this scenic area. Taking advantage of this calm water, outfitters provide night cruises, jet boat trips and self-guided canoe tours. Dead Horse Point State Park, 2,000 feet above the River, provides an outstanding panoramic view of the nearby buttes and canyons, the River below and the La Sal Mountains in the distance. Dead Horse Point, named after a band of mustangs who were innocently left to die of thirst, was the site of the 1978 World Invitational Hang Gliding Tournament.

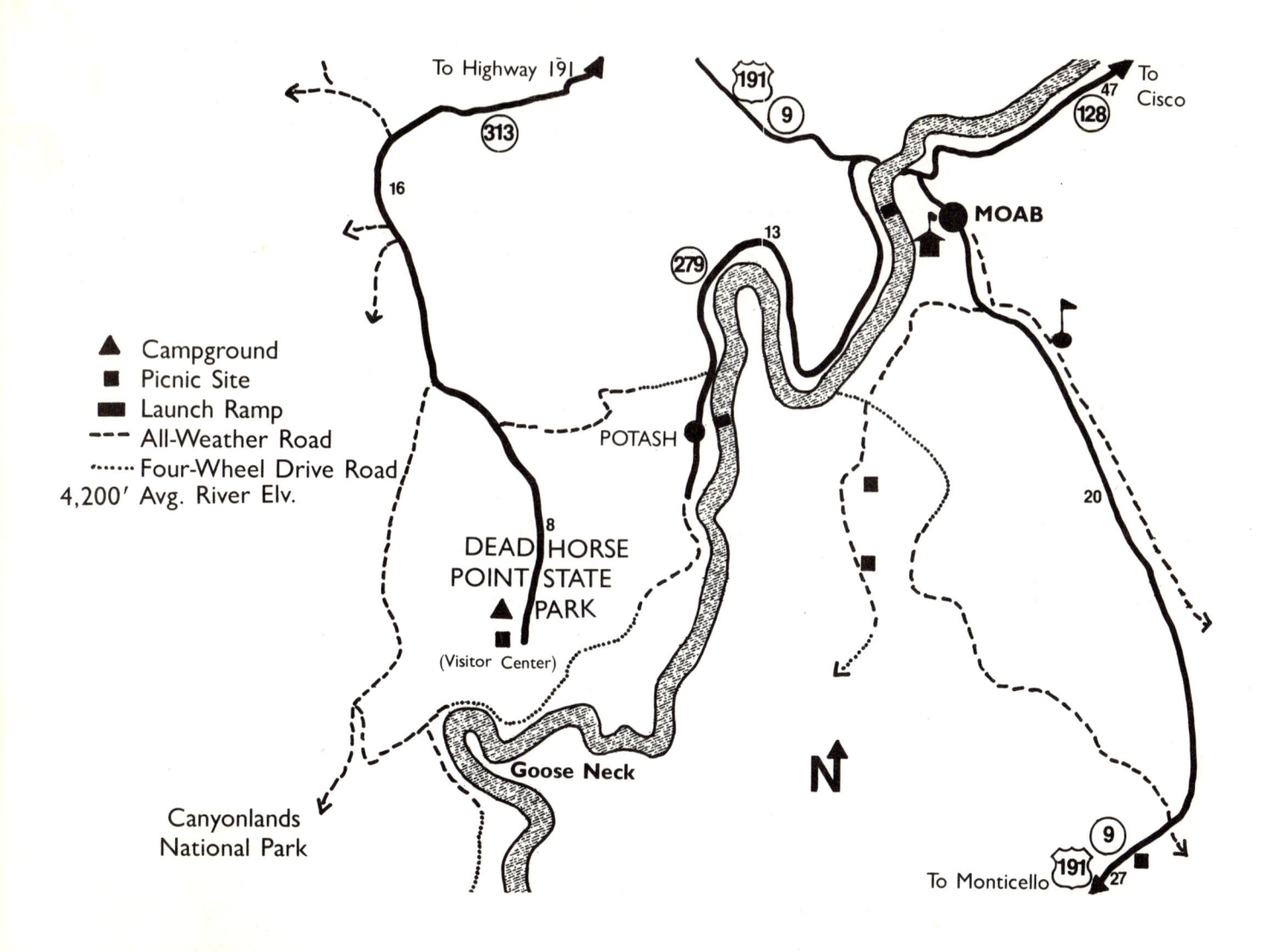

INFORMATION: Bureau of Land Management - Grand Area Office, P.O. Box M, Sand Flat Road, Moab, UT 84532
Ph: 801-259-8193

CAMPGROUNDS & RESORTS	RECREATION	OTHER
Dead Horse Point State Park P.O. Box 609, Moab, UT 84532, Ph: 801-259-6511 21 Tent/R.V. Sites, Electric Hookups, Disposal Station, Visitor Center, Museum	Boating: Raft, Canoe, Kayak and Power Boats Fishing: Channel Catfish Hiking & Horseback Riding Trails Nature & Geological Study 4-Wheeling Jeep, Jet and Float Boat Tours Golf	Major Facilities in Moab

CANYONLANDS NATIONAL PARK

Canyonlands National Park is considered one of the premier primitive Parks in the country. While being relatively inaccessible except by raft, 4-wheeler or on foot, major overlooks are easily reached most of the year by passenger vehicle. Within the Park complex one can view some of the most spectacular geological formations in existence. Facilities within the Park are limited, and water is scarce, so be sure to provide your own drinking water, cooking fuel and supplies. There are many commercial tours available in Moab and Monticello. The quiet flow of both the Green and Colorado Rivers as they journey to the Confluence provides memorable boating opportunities. Once below the Confluence, the Colorado becomes turbulent as it plunges through famed Cataract Canyon. To experience a whitewater raft trip, commercial guides are available and run the River regularly during summer months. In years past, a group called The Friendship Cruise has offered power boaters a unique 184-mile excursion down the Green River to the Confluence and then up the Colorado River to Moab.

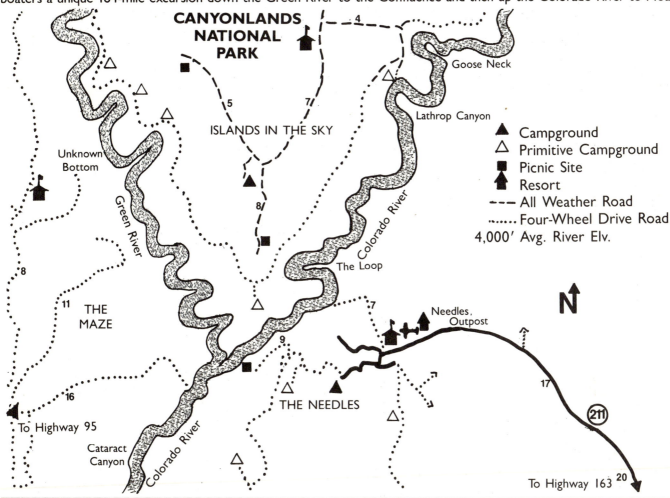

CAMPGROUNDS & RESORTS	RECREATION	OTHER
Canyonlands National Park Islands in the Sky District Willow Flat Campground: 12 Sites, No Water, Pit Toilets, Tables & Firepits Needles District Squaw Flat Campground: 26 Sites, Drinking Water, Pit Toilets, Tables & Firepits 4-Wheel Drive Sites	Boating: Raft, Canoe & Kayak Permits Required: Boating Unit Coordinator, Canyonlands National Park, 125 W. 200 S, Moab 84532, Ph: 801-259-7164 Fishing: Channel Catfish Hiking & Horseback Riding Trails Nature & Geological Study 4-Wheeling, Jeep, Floatboat & Air Tours	**Needles Outpost** Box 1107, Monticello, UT, Ph: 801-259-2032, Tent/R.V. Sites, Hookups, Motel, Store, Snack Bar, Gas, Jeep Rentals, Horseback Trips, Scenic Flights, 4,800 Ft. Airstrip, Propane, Mtn. Bike Rentals **The Friendship Cruise** P.O. Box 4, Green River, UT 94525

INFORMATION: Superintendent, Canyonlands National Park, Moab, UT 84532, Ph: 801-259-7164

CANYONLANDS NATIONAL PARK
By Lin Ottinger

Canyonlands National Park—337,570 acres, encompasses a vast region of dramatic geology, unseen anywhere else on this earth. Contained within this area are awe inspiring arches, spires, windswept mesas and buttes. Elevations range from 4,000 to 8,000 feet, a measure of time and its awesome effect on this fragile land. The area is well named due to the fact that water-eroded-canyons are as common as pawns on a chessboard. This area owes in part its existence to the Colorado River, which is the main receptor of all water caused erosion, an ever changing landscape controlled by the whims of nature.

Be forewarned, the canyonlands is a very primitive area not to be taken lightly. If driving, most of the roads are dirt and not very well maintained, always be sure you carry plenty of drinking water. It is advisable to have a good map or a trained guide, whether driving or backpacking. The park boundary is not shown on this map as it has no meaning so far as scenic values are concerned.

This map was published by Lin Ottinger, artwork by Jean-Claude Gal and is available printed on poster quality paper 18 x 22 inches at:

Lin Ottinger Tours
137 N. Main Street
Moab, Utah 84532

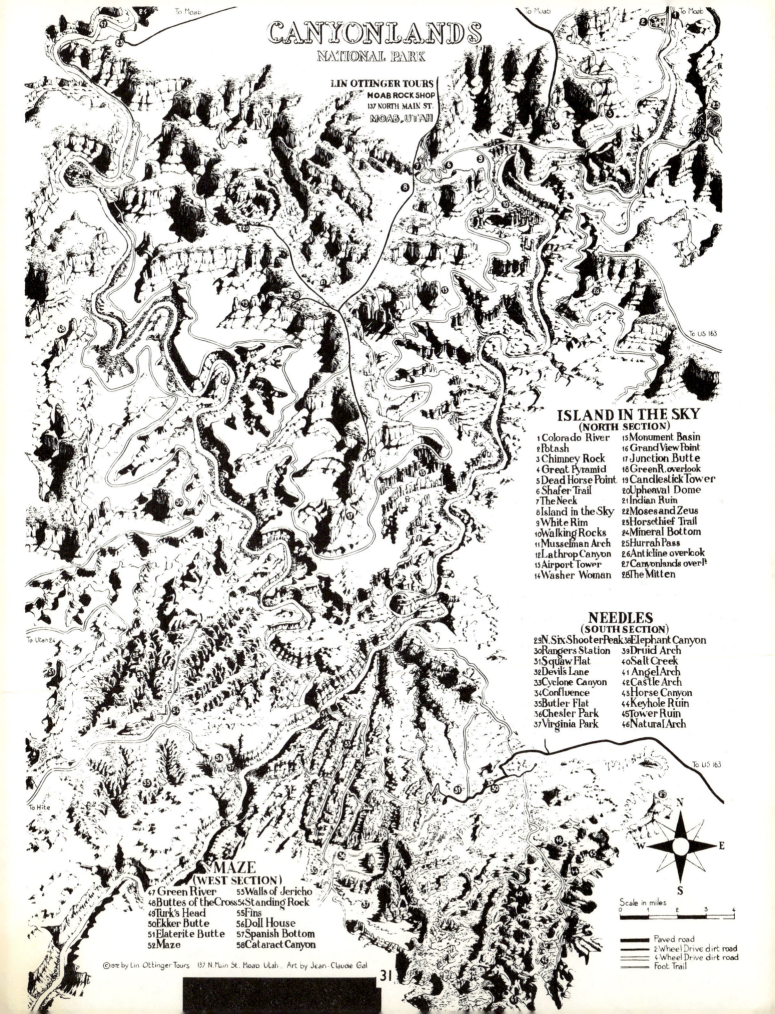

CANYONLANDS
NATIONAL PARK

LIN OTTINGER TOURS
MOAB ROCK SHOP
137 NORTH MAIN ST.
MOAB, UTAH

To Moab

To Hite

To Utah 24

To US 163

ISLAND IN THE SKY
(NORTH SECTION)

1 Colorado River
2 Potash
3 Chimney Rock
4 Great Pyramid
5 Dead Horse Point
6 Shafer Trail
7 The Neck
8 Island in the Sky
9 White Rim
10 Walking Rocks
11 Musselman Arch
12 Lathrop Canyon
13 Airport Tower
14 Washer Woman
15 Monument Basin
16 Grand View Point
17 Junction Butte
18 Green R. overlook
19 Candlestick Tower
20 Upheaval Dome
21 Indian Ruin
22 Moses and Zeus
23 Horsethief Trail
24 Mineral Bottom
25 Hurrah Pass
26 Anticline overlook
27 Canyonlands overl?
28 The Mitten

NEEDLES
(SOUTH SECTION)

29 N. Six Shooter Peak
30 Rangers Station
31 Squaw Flat
32 Devil's Lane
33 Cyclone Canyon
34 Confluence
35 Butler Flat
36 Chesler Park
37 Virginia Park
38 Elephant Canyon
39 Druid Arch
40 Salt Creek
41 Angel Arch
42 Castle Arch
43 Horse Canyon
44 Keyhole Ruin
45 Tower Ruin
46 Natural Arch

MAZE
(WEST SECTION)

47 Green River
48 Buttes of the Cross
49 Turk's Head
50 Ekker Butte
51 Elaterite Butte
52 Maze
53 Walls of Jericho
54 Standing Rock
55 Fins
56 Doll House
57 Spanish Bottom
58 Cataract Canyon

Scale in miles
0 1 2 3 4

Paved road
2 Wheel Drive dirt road
4 Wheel Drive dirt road
Foot Trail

N
W E
S

ROCKHOUND GEOLOGY
By Lin Ottinger

These are some of the pages of history of the earth's crust which are exposed near Moab, Utah. One or more of these layers might be encountered anywhere throughout the intermountain area on the Colorado Plateau. These layers of sediments tell the story of what creatures lived and what events took place in eons past, making this area a geologic wonderland. Learn to read them and a new book will be opened to you.

All these sedimentary formations were created when wind, water and the fingers of time wore away high mountains of igneous and previously formed and uplifted sedimentary rock; the sediment was deposited in the lowlands and the seas which at various times covered this area.

After the last uplift of the Colorado Plateau occurred, the Colorado and other rivers cut through the sedimentary layers leaving canyons thousands of feet deep. Numerous faults broke these sedimentary layers leaving the same layer exposed at river level and on the highest plateaus. The anticlines, synclines and grabens that are abundant in the Moab area aided the exposure of many of these formations making them available for you to study and enjoy. Learning to recognize formations that contain gems, minerals and fossils puts the entire Colorado Plateau in the palm of your hand. Wherever you encounter these formations you are likely to find your favorite specimens.

The brief, but heavy rains and arid climate of the area support little vegetation that would hold the overburden. This leaves the formations eroded clean and easy to study. Gems, minerals and fossils are so abundant that it is possible to find them in new locations or even varieties not previously discovered.

This map was published by Lin Ottinger, artwork by Jean-Claude Gal and is available printed on poster quality paper 18 x 22 inches at:

Lin Ottinger Tours
Moab Rock Shop
137 N. Main Street
Moab, Utah 84532

ROCKHOUND GEOLOGY MAP

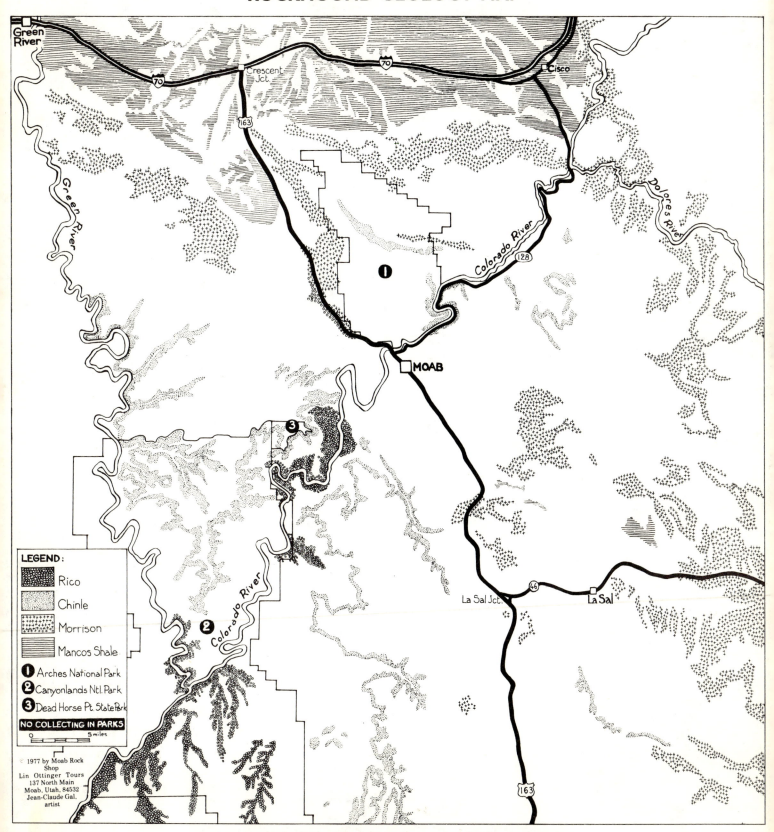

LEGEND:

- Rico
- Chinle
- Morrison
- Mancos Shale
- ❶ Arches National Park
- ❷ Canyonlands Ntl. Park
- ❸ Dead Horse Pt. State Park

NO COLLECTING IN PARKS

0   5 miles

© 1977 by Moab Rock
Shop
Lin Ottinger Tours
137 North Main
Moab, Utah, 84532
Jean-Claude Gal,
artist

This map shows you the exposures of four formations, Mancos Shale, Morrison, Chinle, and Rico, which are most interesting to the rockhound. Data was taken from the U.S. Geological Survey.

Use this map and text in conjunction with your road map or topographic map to find the roads into outcrops of the various formations.

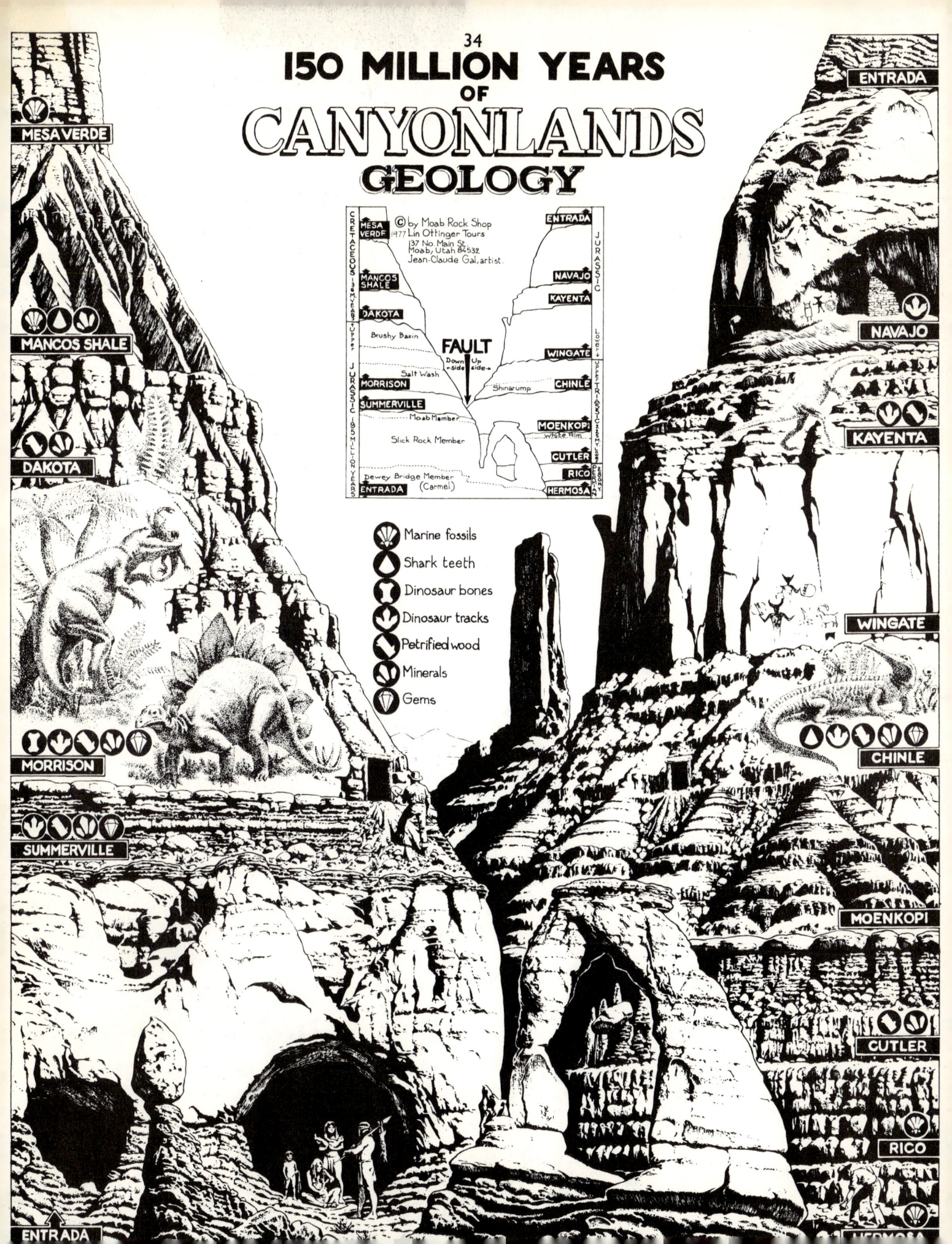

GLEN CANYON NATIONAL RECREATION AREA
LAKE POWELL

Glen Canyon is the largest canyon system on the Colorado River. For centuries, this mighty River and its tributaries carved a maze of narrow twisting canyons through the slickrock desert. A geological history of former oceans, swamps and deserts are reflected in this harsh landscape. Colorful red and buff mountains, towers, cliffs, buttes, arches and bridges present an unparalleled scene of sculptured beauty.

Man has been drawn to Glen Canyon for thousands of years by its great Rivers and natural beauty. Evidence of prehistoric Indians, known as the Desert Archaic, is found in chipped stone tools and weapons, grinding stones and baskets. Amateur and professional archaeologists find bits of ancient pottery, mysterious petroglyphs and Indian Kivas. The Anasazi, Paiute, Navajo and Pueblo Indians have made their historic mark. Spanish, Mexican and American adventurers were drawn to this remote land. While man had made his claim, his presence prior to 1963 had relatively nominal effect.

The Glen Canyon Dam began storing water in 1963. The resulting Lake was named after John Wesley Powell. Powell was a true American hero. After losing an arm in the Civil War, this famed adventurer explored and mapped this rugged canyon. Later, he helped create the Bureau of Reclamation. His philosophy of balancing man's need with the environment led to the conservation movement. Lake Powell, reflecting this balance, is a living monument to this great American.

Clear azure blue water of Lake Powell contrasts with the surrounding landscape creating a scene of breathtaking beauty. Where once only the hardy were able to reach this remote area, thousands are now able to easily view its 96 major canyons. The 1,950 miles of shoreline larger than the Western Coast of the United States, are still relatively unreachable by road, but by water, this shoreline is easily accessible. When full, the Lake reaches a depth of 568 feet and covers a surface of 252 square miles.

Lake Powell is now one of America's premier year-round recreation attractions. This is boating country! Canoes, kayaks and other small craft are advantageous in the smaller narrow canyons. Afternoon winds attract the sailor, and the wide channels and bays are perfect for the waterskiier. Houseboats are abundant, and boat camping anywhere along the shoreline is popular. Numerous beaches and sandy coves await the swimmer and the crystal clear water offers the diver a unique view of the submerged landscape. The fisherman will find twelve varieties of game fish. Hiking, backpacking and 4-wheeling in the arid region around the Lake is popular.

The Glen Canyon Recreation Area is administered by the National Park Service. Park Rangers patrol the Lake by boat and plane, and offer advice and assistance as well as enforcing regulations. Navigational aids supplied by the U.S. Coast Guard and maintained by the Park Service are subject to change. Contact a Ranger Station for current status and conditions. While boating, be sure to have enough fuel. Gas is available only every 50 miles at Marinas on this huge Lake.

For more specific information, see the following pages.
Detailed Navigational Maps are sold at Marinas.

GLEN CANYON
NATIONAL RECREATION AREA
LAKE POWELL

207 MILES OF RIVER SHOWN

▲ Campground
♟ Resort
★ Marina
▭ Boat Dock
⋯⋯ Four Wheel Drive Road
3,700' Avg. River Elev.

TO HANKSVILLE

95
30

Cataract Canyon

30

20
95

NOTAM ROAD

276

Hite Marina
Airport-Emergency Use Only

36
95

Natural Bridges National Monument

41
5

8

TO BLANDING

Good Hope Bay

Bullfrog Marina

19

ESCALANTE

12
6

54

Escalante River

Halls Crossing Marina

14

9

263

Clay Hills Crossings

10

23

RIVER

JUAN

SAN

Last Chance Bay

Dangling Rope Marina

Rainbow Bridge National Monument

6

UTAH
ARIZONA

89
Wahweap Marina

Lees Ferry

PAGE

39
150 To Flagstaff

Navajo Canyon

98
51 To Hwy 160

Navajo Indian Reservation

N

Excellent Publications on Lake Powell:

Fishing Lake Powell by Bob Hirsh and Stan Jones

Stan Jones' Boating and Exploring Map

Sun Country Publications
P.O. Box 955
Page, Arizona 86040

Dowlers' Lake Powell Boat & Tour Guide by The Warren L. Dowlers

The Warren L. Dowlers
526 Camillo Street
Sierra Madre, California 91024

Glen Canyon—Lake Powell—The Story Behind the Scenery by Ronald E. Everhart

KC Publications
Bow 14883
Las Vegas, Nevada 89114

INFORMATION: Glen Canyon National Recreation Area, Superintendent, Box 1507, Page, Arizona 806040, Ph: 602-645-2511

CAMPGROUNDS & RESORTS	RECREATION	OTHER
See Following Pages for Descriptions Disposal Stations at All Marinas for Boats & R.V.s	Boating: Open to All Boating, Water Skiing & Boat Camping	Park Service 24 Hours per Day Emergency Phone: 602-645-9585
Reservations for Houseboats, Power-boats & Accommodations	Fishing: Large & Smallmouth & Striped Bass, Bluegill, Catfish, Green Sunfish, Northern Pike, Walleye, Rainbow & Brown Trout	
Del E. Webb Recreational Properties Box 29040 Phoenix AZ 85038 Ph: Toll Free 1-800-528-6154 or 602-278-8888	Full Service Marinas Launch Ramps & Boat Rentals Tours & Storage Swimming & Scuba Diving Hiking & Backpacking 4-Wheel Trails & Jeep Tours	

GLEN CANYON NATIONAL RECREATION AREA
CATARACT CANYON TO
HITE RESORT & MARINA

Just below the confluence of the Green River, the Colorado again asserts its power. Flowing past Needles Country of Canyonlands National Park into Glen Canyon National Recreation Area, Cataract Canyon provides float boaters outstanding adventure. Cataract Canyon is one of North America's most challenging whitewaters so its Class IV rapids are not for the amateur. There are many commercial tours running this famed section of the Colorado. After leaving the notorious "Big Drop" rapids, the River slows to a leisurely pace into the upper canyons of Lake Powell. Spent rafters merge with power boaters and houseboats near Hite Marina. There is excellent catfishing at the mouth of Dirty Devil River and North Wash.

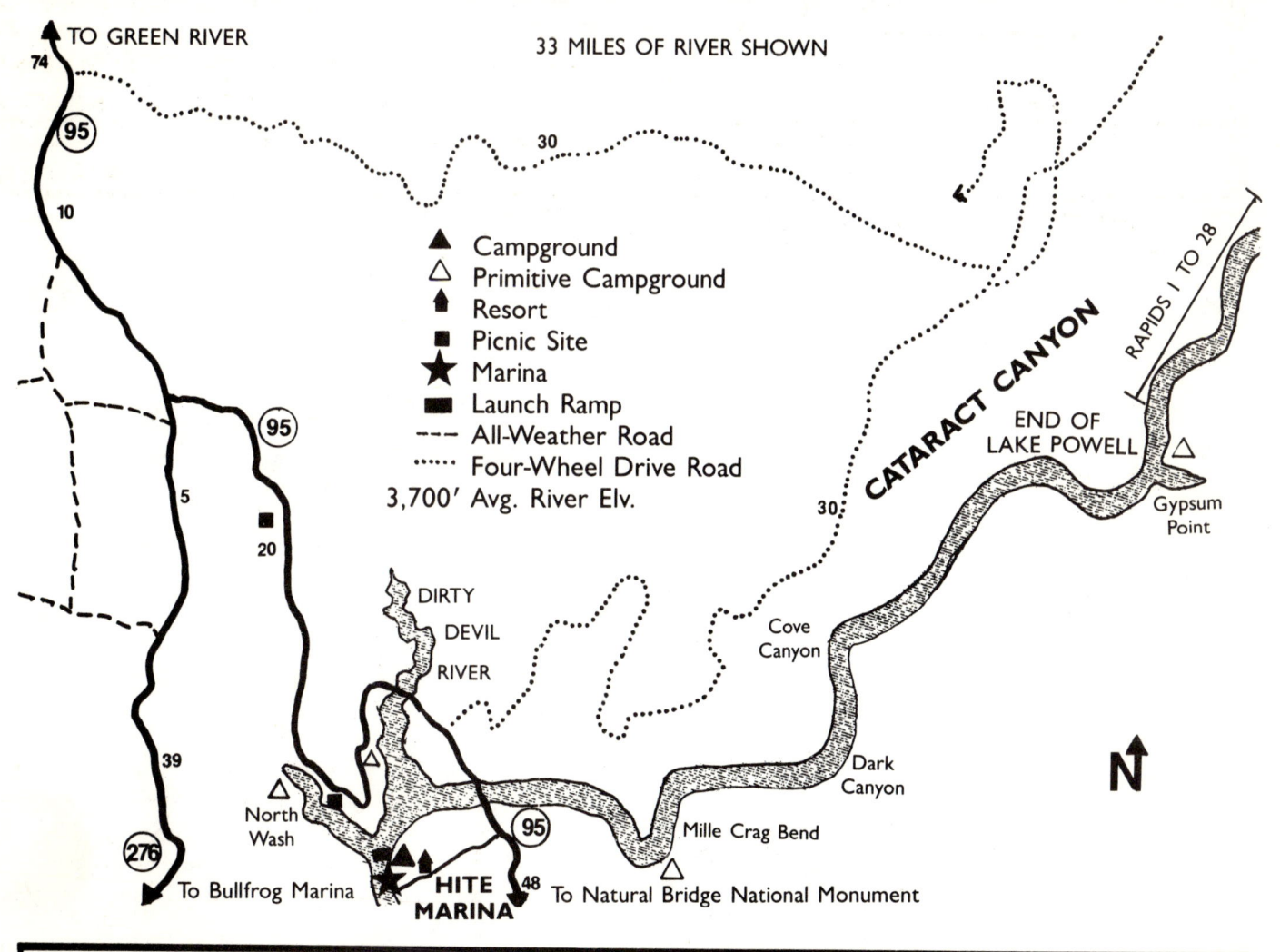

INFORMATION: Hite Resort & Marina, Hanksville, UT 84734, Ph: 1-800-528-6154; Arizona 602-278-8888, Utah 801-684-2278

CAMPGROUNDS & RESORTS	RECREATION	OTHER
Hite Resort & Marina—Graded Campsite Area, 5 Housekeeping Units, Grocery Store, Gas Station, Camping Supplies, Bait & Tackle, Disposal Station for Boats & R.V.s Open Boat Camping Anywhere on Lakeshore	Houseboat & Powerboat Rentals Whitewater Boating: Cataract Canyon Class IV; Permits Required See Previous Graph for Further Information	Boating Unit Coordinator Canyonlands National Park 125 W. 2nd South Moab, UT 84532 Ph: 801-259-7164

GLEN CANYON NATIONAL RECREATION AREA
BULLFROG AND HALLS CROSSING RESORTS AND MARINAS

Imagine waterskiing on Good Hope Bay surrounded by colorful rising rock cliffs. Boat past the largest rock island in the Lake, Bass Rock, or relax at a quiet campsite on the shore of Red Canyon. Good Hope Bay and Red Canyon are said to have sizeable walleye populations while crappie are abundant in upper Hansen Creek. Largemouth bass are plentiful at Halls Creek and Bullfrog Bays. There are excellent campsites at Halls Creek, Bullfrog and Farley Canyon. Indian artifacts, ruins and petroglyphs are found in Moki, Forgotten and White Canyons. These are but a few of the many attractive opportunities found in this section of Lake Powell. The new ferry in service between Bullfrog and Halls Crossing will save the driver 110 miles.

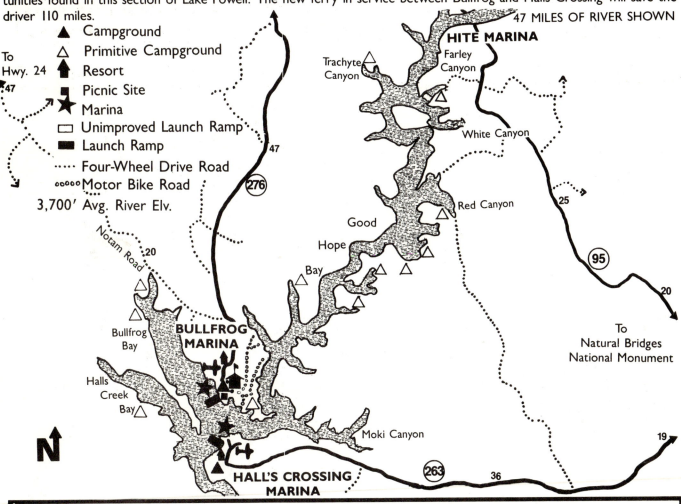

▲ Campground
△ Primitive Campground
⬆ Resort
▪ Picnic Site
★ Marina
▭ Unimproved Launch Ramp
▬ Launch Ramp
···· Four-Wheel Drive Road
ooooo Motor Bike Road
3,700' Avg. River Elv.

47 MILES OF RIVER SHOWN

INFORMATION: Bullfrog Resort & Marina, Hanksville, UT 84734, Ph: 801-684-2233
Halls' Crossing Resort & Marina, Blanding, UT 84511, Ph: 801-684-2261

CAMPGROUNDS & RESORTS

Bullfrog Resort & Marina—R.V. Park, Full Hookups, 40 Housekeeping Units, Motel, Lodge, Restaurant & Lounge, Gift Shop, Store, Full Service Gas Station, 3,500 Ft. Paved Airstrip

Hall's Crossing Resort & Marina—32 R.V. Sites, Full Hookups, 16 Housekeeping Units, Full Service Gas Station, Store, Snack Bar, Marina Store, 3,685 Ft. Graded Airstrip

RECREATION

Houseboat & Powerboat Rentals
Lake Tours
Fishing Guides
Motor Bike Trails

See Previous General Graph for Further Information

OTHER

Bullfrog:
National Park Service
Ranger Office
Ph: 801-684-2212
Improved Campground, Picnic Area, Disposal Station, Paved Launch Ramp

Hall's Crossing:
National Park Service
Ranger Office
Ph: 801-684-2270
Improved Campground, Disposal Station, Paved Launch Ramp

GLEN CANYON NATIONAL RECREATION AREA
RAINBOW BRIDGE AND DANGLING ROPE MARINA

While facilities are somewhat limited in this section of Lake Powell, there is a boat-in Marina at Dangling Rope with supplies and fuel. Stock up and enjoy an extended cruise into the arms of the Escalante and San Juan Rivers. These Rivers provide a variety of geological formations, Indian ruins and recreational opportunities in their varied and remote Canyons. Historic Hole-In-The-Rock was created by early Morman road builders on their way to the San Juan. Rainbow Bridge, the world's largest natural bridge, is now easily accessible by boat and a short walk from the courtesy dock. Swim, ski, hike, fish or just relax in this beautiful area. Be sure to bring a camera.

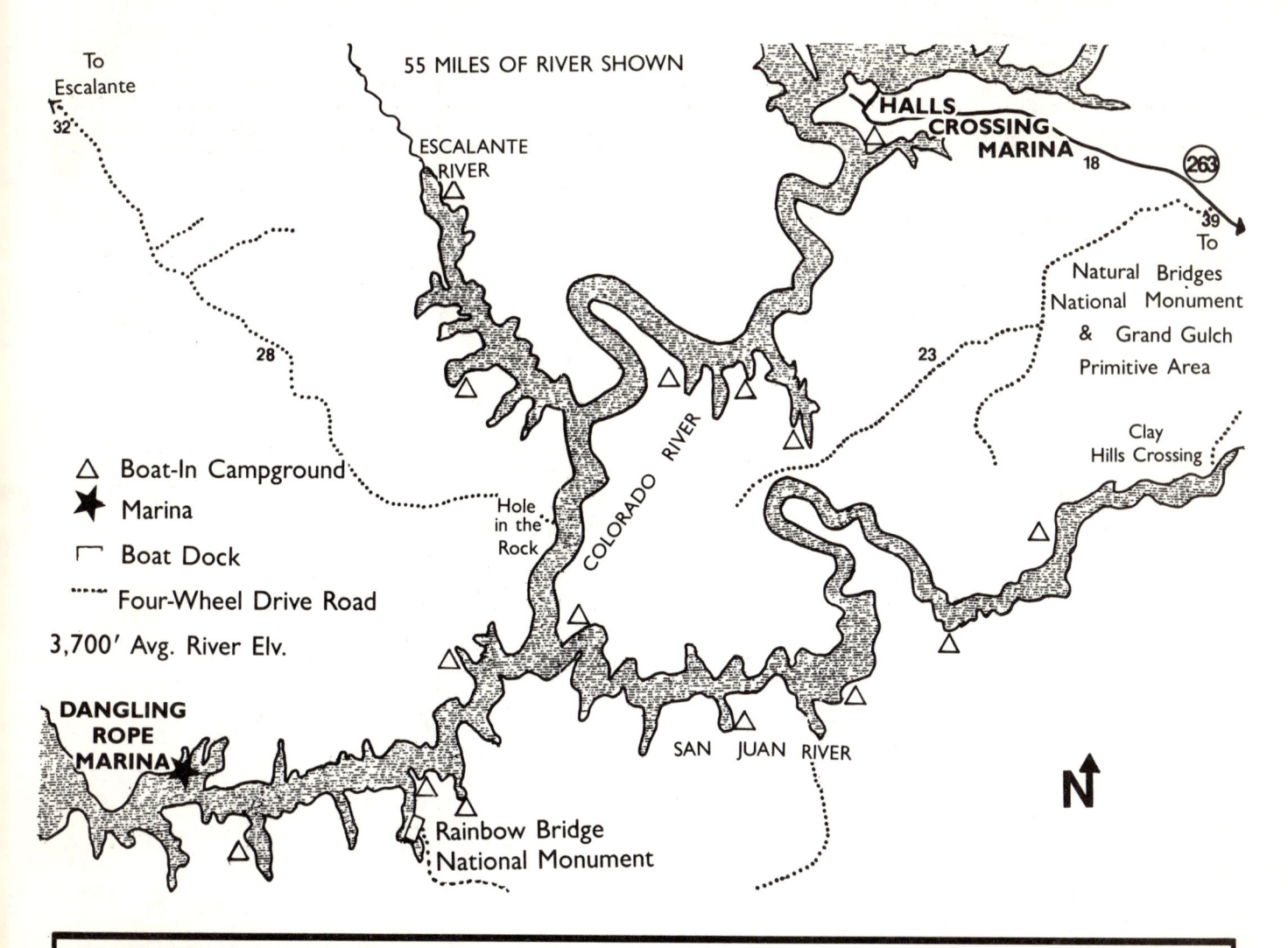

INFORMATION: Dangling Rope Marina, Call Wahweap Lodge & Marina for Information: Ph: 602-645-2433		
CAMPGROUNDS & RESORTS	**RECREATION**	**OTHER**
Accessible Only by Boat—Fresh Water, Disposal Station, Ranger Station, Full Service Marina, Fuel, Docks, Store, Groceries, Bait & Tackle	**San Juan River:** Montezuma Creek— Clay Hills Crossing: Rafting: Class II — Permit Required — Not Runnable in Late Summer and Fall During Dry Years	For Information Regarding San Juan River to Clay Hills Crossing and the Adjacent Grand Gulch Primitive Area, Contact: **Bureau of Land Management:** San Juan Resource Area P.O. Box 7 Monticello, UT 84535 Ph: 801-587-2201

GLEN CANYON NATIONAL RECREATION AREA
WAHWEAP BAY AND GLEN CANYON DAM

This southern section of Lake Powell offers an abundance of accommodations, campsites and marine facilities. Enjoy a romantic evening dinner cruise aboard the Canyon King Paddlewheeler. Beautiful Padre Bay, the Lake's widest, and Last Chance Bay, offer good fishing and waterskiing. Navajo Canyon's wide twisting waterway leads into historic Anasazi Country, now sparsely settled by the Navajo. Warm Creek and Wahweap Bays are popular boating, fishing and waterskiing areas. Just below Wahweap Bay, the Glen Canyon Dam rises 583 feet above bedrock. The Carl Hayden Visitor Center presents an audiovisual show, exhibits, a relief map, self-guided dam and power plant tours.

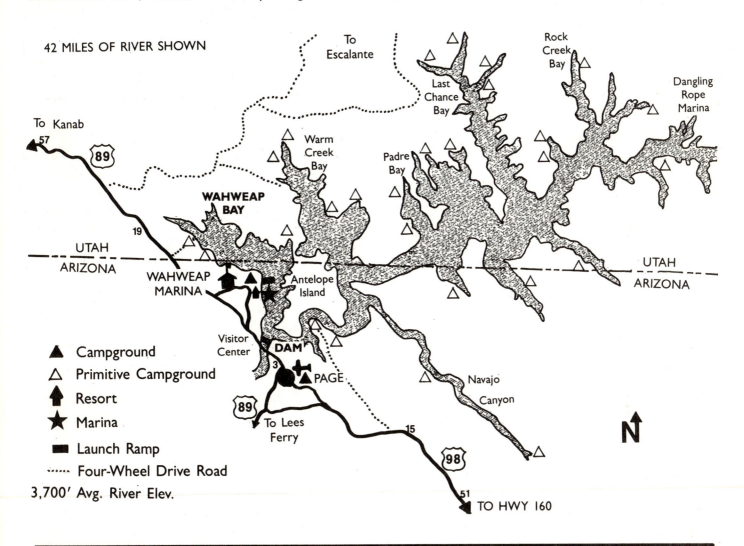

INFORMATION: Wahweap Lodge and Marina, P.O. Box 1597, Page, AZ 86040, Ph: 602-645-2433		
CAMPGROUNDS & RESORTS	**RECREATION**	**OTHER**
Wahweap Lodge and Marina— 118 Tent/R.V. Sites, Full Hookups, Laundry, Picnic Area, Playground, General Store, Lodge, Swimming Pool, 2 Restaurants, Lounge and Gift Shop, Paddlewheel, Evening Dinner and Other Lake Cruises, Gas Station, Courtesy Transportation to Private Airport Facility, Rental Cars	Houseboat and Powerboat Rentals **Wahweap Campground** National Park Service Ph: 602-645-8883 180 Tent/R.V. Sites, Laundry, Groceries, Swim Beach, Fish Cleaning Station, Recreation Program, Visitor Center, Ranger Station	**Page-Lake Powell Campground and Trailer Motel** P.O. Box BB, Page 86040 Ph: 602-645-3374 150 Tent/R.V. Sites, Full Hookups, Disposal Station, Laundry, Groceries, Propane, Trailer Rental, Pets O.K., Nightly and Weekly Fishing License and Tackle

GLEN CANYON NATIONAL RECREATION AREA
GLEN CANYON DAM TO LEE'S FERRY

The cold clear water flowing from the bottom of Glen Canyon Dam again becomes the Colorado River. It seems to be gathering strength for its encounter with the angry rapids waiting fifteen miles below. Passing through the sheer, narrow towering walls of Glen Canyon, this relatively inaccessible section of the River provides a scenic environment for the boater, camper and angler. This is the home of huge trout. A popular one day float trip leaves the Powell Museum daily. Boaters are cautioned to stay upriver from the launch ramp at Lee's Ferry. Dangerous rapids are below the cable, and only those with permits are allowed. Historic Lee's Ferry is the dividing line between the Upper and Lower Colorado River Basins.

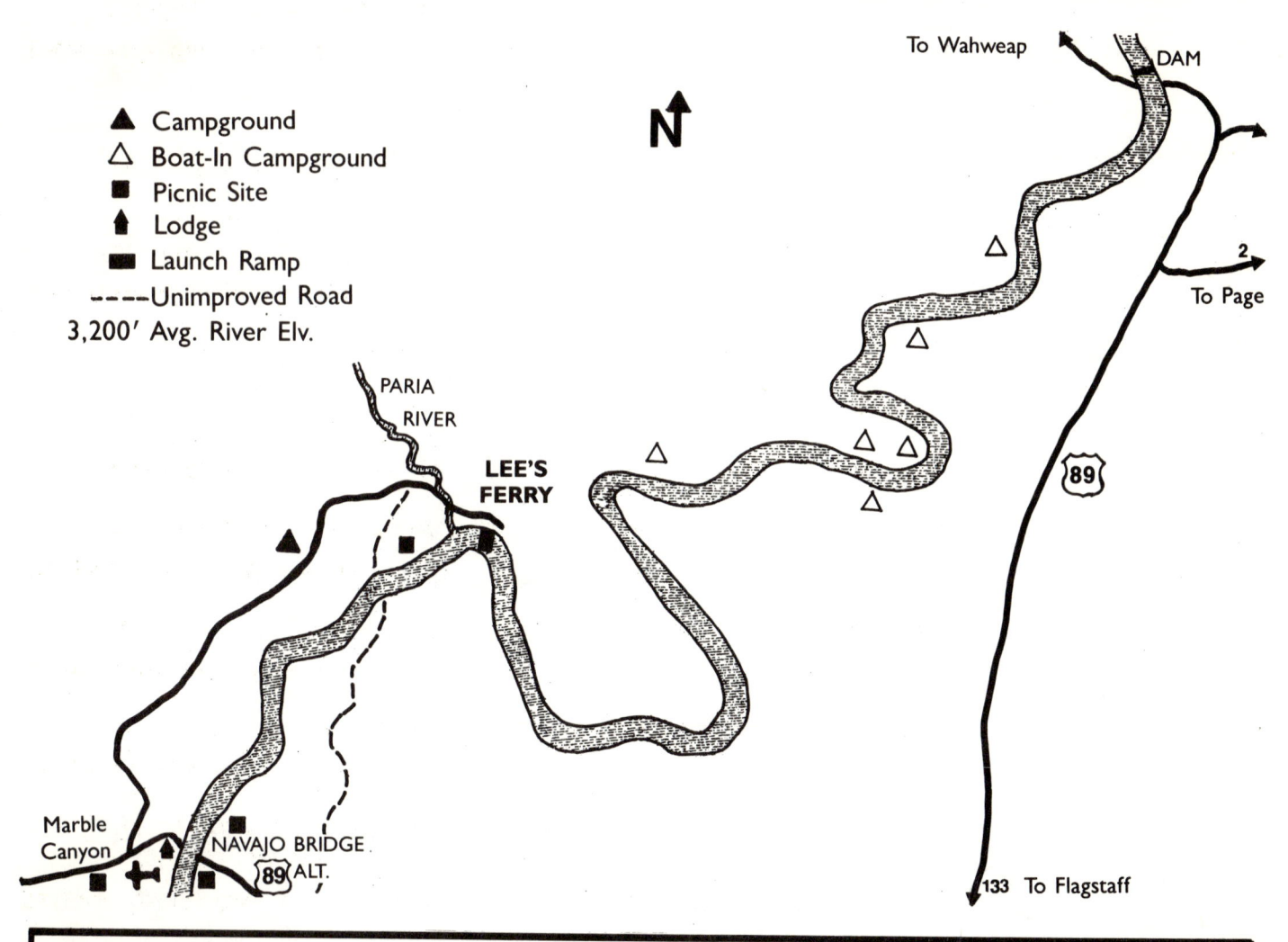

- ▲ Campground
- △ Boat-In Campground
- ■ Picnic Site
- ♦ Lodge
- ▬ Launch Ramp
- ---- Unimproved Road

3,200' Avg. River Elv.

INFORMATION: Glen Canyon National Recreation Area, P.O. Box 1507, Page, AZ 86040, Ph: 602-355-2234

CAMPGROUNDS & RESORTS	RECREATION	OTHER
Lee's Ferry Campground—See Above for Information. 56 Tent/R.V. Sites, Launch Ramp, 6 Upriver Boat Camps, 18 Sites **Marble Canyon Lodge**—Highway 89, Marble Canyon 86036, Ph: 602-355-2225. Lodge, Restaurant, Gas Station & Store, 4,000 Ft. Airstrip Adjacent to Lodge	Boating: Upriver to 1,000 Ft. Below Dam. No Boating Without Permits Below Launch Ramp. **Dangerous** Rapids Downstream from Cable Fishing: Rainbow, Brook, Brown & Cutthroat Trout Float Trips: Dam to Lee's Ferry— May through September Air Tours, Picnic Area, Nature Walk, Backpacking, Horse Rentals	Full Facilities in Page Area Page-Lake Powell Chamber of Commerce P.O. Box 727, Page 86040 Ph: 602-645-2741 John Wesley Powell Museum 6 North Lake Powell Blvd. Page, AZ 86040

Boat Tours
Lake Powell — Glen Canyon National Recreation Area

Group Charters — From Bullfrog, Hall's Crossing and Hite Marinas

Individuals or Small Groups: Guided Boat Tours
Ph: Toll Free — 800-528-6154 or 602-278-8888

Rainbow Bridge Cruise — From Bullfrog, Hall's Crossing and Wahweap Marinas

All Day 100-Mile Round Trip Includes Box Lunch
Half Day from Wahweap Only

Defiance House — From Bullfrog Marina

2½ Hour Cruise to Forgotten Canyon

Navajo Tapestry — From Wahweap Marina

One Hour Cruise to Glen Canyon Dam via Antelope Canyon and Navajo Canyon

Wahweap Bay Paddlewheeler

One Hour Cruise on an Authentic 95-foot Paddlewheel Riverboat

Sunset/Moonlight Cruise Including Cocktails & Buffet Dinner
Groups Accommodated with Advance Reservations

_____

Wilderness River Adventures

P.O. Box 717, Page, AZ 86040
Ph: 602-645-3296

All Day Raft Trips from Page

_____

Lee's Ferry Guide Service

P.O. Box 2702, Page, AZ 86040
Ph: 602-645-5127

For Trout Fishing: 20 ft. Sea Ox Fishing Boat with Guide
Fishing Gear Rental Available

Scenic Flights

Lake Powell Air Service
P.O. Box 1385
Page Airport
Page, AZ 86040
Ph: 602-645-2494

Rainbow Bridge	½ Hour
Escalante River	1 Hour
Bryce Canyon	1½ Hours
The Grand Canyon	1½ Hours
Monument Valley	1½ Hours
Extended Grand Canyon	2 Hours
Canyonlands	2½ Hours
Monument Valley-Navajo Nation	4 Hours
Eagle Express	8 Hours

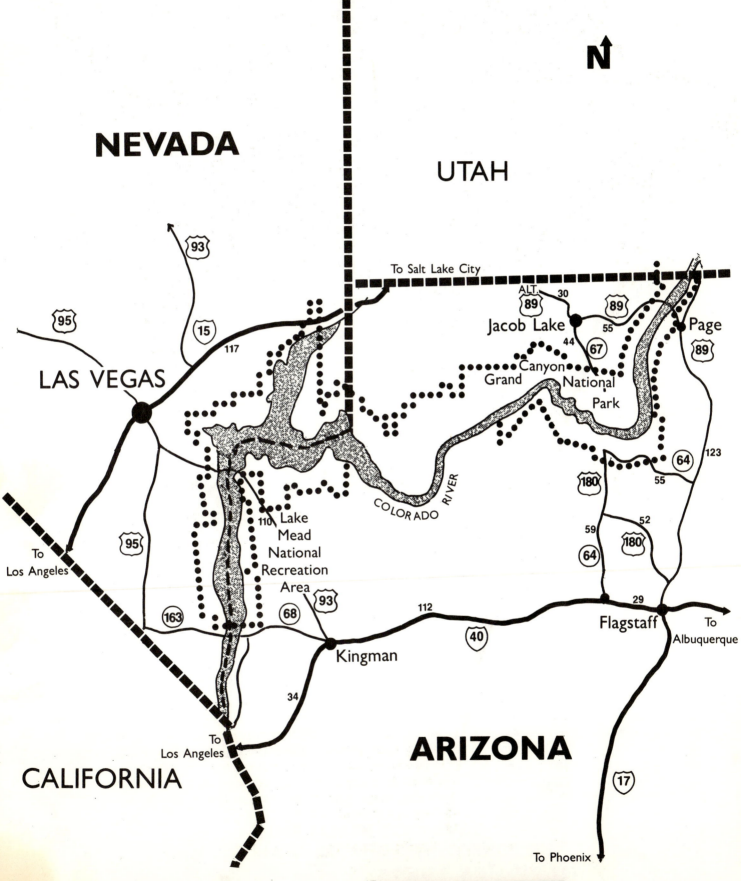

N

NEVADA

UTAH

93

To Salt Lake City

95

15

117

ALT. 89   30   89

Jacob Lake   55

44   67

LAS VEGAS

Canyon   National

Grand   Park

Page

89

COLORADO RIVER

64   123

180   55

110   Lake
Mead
National
Recreation
Area

52

59   180

64

To
Los Angeles

95

112

29

163   68   93

40

Flagstaff   To
Albuquerque

Kingman

34

To
Los Angeles   ARIZONA

17

CALIFORNIA

To Phoenix

LEE'S FERRY TO THE CONFLUENCE OF THE LITTLE COLORADO

Lee's Ferry is the start of one of the most beautiful and exciting River adventures in the world. Quietly entering the longest and deepest canyon carved by the Colorado at the confluence of the Paria River, the River begins its 277.4 mile journey through the majesty of the Grand Canyon. Dropping from an elevation of 3,107 feet at Lee's Ferry to Lake Mead's elevation of 1,229 feet, the River flows through 161 named rapids. There are 26 major rapids ranging from 5 to 10 depending on water flow. In order to preserve the natural environment, the National Park Service has limited river traffic. Commercial outfitters, licensed by the Park Service, comprise the majority of "river runners." Advance reservations are advised.

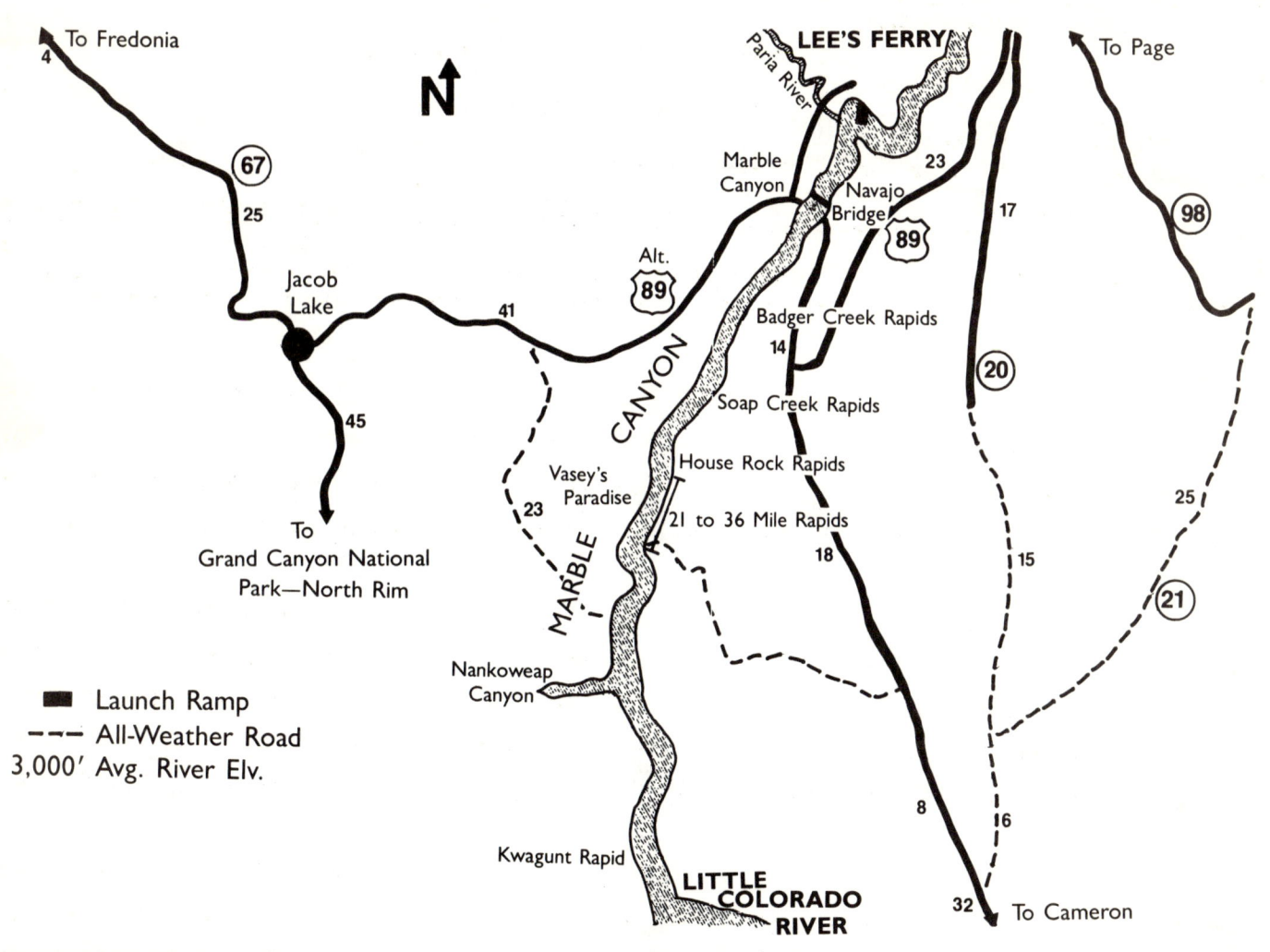

INFORMATION: River Permits Office, Grand Canyon National Park, P.O. Box 129, Grand Canyon, Arizona 86023 Ph: 602-638-7843		
CAMPGROUNDS & RESORTS	**RECREATION**	**OTHER**
See Previous Pages	Whitewater Boating: Rapids Class 2–10—Permits Required in Advance Commercial River Trips by Motorized or Oar Powered Rafts, Dorys and Kayak Support Fishing: Rainbow, Brown Trout, Crappie and Bluegill Hiking & Backpacking Nature Study	Full Facilities in Page Resort & Supplies at Marble Canyon Lodge

GRAND CANYON
NORTH AND SOUTH RIMS
INNER CANYON

The Colorado now enters the heart of the Grand Canyon. Flowing one mile below the North and South Rims, the River carves its way through an unsurpassed display of geological formations. This majestic stage for nature's power and dramatic moods attracts over 3 million visitors from all over the world each year. The scenery is awesome.

The South Rim's Grand Canyon Village and nearby town of Tusayan offers an abundance of facilities, guided tours, visitor's information and Park Headquarters. The South Rim is the hub of the Park attracting the majority of visitors. A leisurely walk along part or all of the South Rim Nature Trail is popular, and the driver is rewarded along the East and West Rim Drives by a number of scenic vistas.

Although only 11 miles separate the North and South Rims, it is a 214 mile drive between the two Rims. The North Rim, closed during the winter, offers a relatively remote contrast to the busy South Rim. A gas station, general store, campground and Ranger Station are located at the Village, and the Grand Canyon Lodge is located one mile from Bright Angel Point. The Bright Angel Nature Trail provides an interesting panorama of the Canyon below. Both Rims provide a number of spectacular views of this gorge.

A 20.8 mile foot trail crossing the River at the Kaibab Suspension Bridge, connects both Rims. The Inner Canyon is only for the experienced hiker in good physical condition. There are mule trips for those who wish to enjoy this remote area. The rugged backcountry offers developed, primitive and undeveloped camping and the facilities at Phantom Ranch. Permits and reservations are required.

INFORMATION: Superintendent. Grand Canyon National Park, P.O. Box 129, Grand Canyon, AZ 86023, Ph: 602-638-7888

CAMPGROUNDS & RESORTS	RECREATION *	OTHER
See Following Pages	Whitewater Boating: See Previous Pages Commercial River, Air, Bus Trips, Trail Guides Hiking & Backpacking: See Following Pages Fishing: River & Streams—Rainbow, Cutthroat & Native Trout Hunting: Deer—Kaibab National Forest Picnicking Scenic Walks & Drives Nature & Geological Study Indian Artifacts	**Grand Canyon Chamber of Commerce**— P.O. Box 235, Williams, AZ 86046 Ph: 602-635-2041 **Kaibab National Forest** 800 S. 6th St., Williams, AZ 86046 Ph: 602-635-2681

NORTH RIM

The North Rim is generally open mid-May through late-October depending on the weather.

Campgrounds

North Rim: 82 Tent/R.V. Sites, No Hookups, Disposal Station, Limited Groceries and Supplies, Public Showers and Laundry Nearby. No Propane Nearby.

Group Reservations: Contact North Rim, Grand Canyon National Park, Box 129, Grand Canyon, AZ 86023 Reservations Ph: 801-586-7686

DeMotte: 20 Tent/R.V. Sites, No Hookups, Groceries.

Contact: National Forest, Ph: 602-643-5895

Jacob Lake: 48 Tent/R.V. Sites, No Hookups, Groceries, Nature Trails.

Jacob Lake R.V. Park: 32 R.V. Sites, Full Hookups.

Lodges

Grand Canyon Lodge: Cabins, Dining Room, Lounge, Curios, Cafeteria.

Contact: TWA Services, Inc., Box 400, Cedar City, Utah 84720, Ph: 801-586-7686

Kaibab Lodge: Modest Rooms and Restaurant, Store and Gas Station Nearby. 18 Miles North of Rim.

Contact: Kaibab Lodge, Fredonia, AZ 86022, Ph: 602-638-2389

Jacob Lake Inn: Motel and Rustic Cabins, Dining Room & Coffee Shop

Contact: Jacob Lake Inn, Jacob Lake, AZ 86022, Ph: 602-643-5532

SOUTH RIM

Campgrounds

Mather: 317 Tent/R.V. Sites, No Hookups, Disposal Station, Laundry, Store, Showers, Restaurant, Gas. Reservations Required May 15 through September 30 by Mail or Ticketron. Contact National Park for Group Reservations.

Desert View: 50 Tent/R.V. Sites, No Hookups, General Store, Gas Station, Gift Shop. Open Mid-May through Mid-October. 26 Miles East of Village.

For Information Contact: Grand Canyon National Park, Box 129, Grand Canyon, AZ 86023. Ph: 602-638-7888

Ten X: 70 Tent/R.V. Sites, No Hookups, 10½ Miles South on SR 64. Operated by the Kaibab National Forest.

Trailer Village: 84 R.V. Sites, Full Hookups, Disposal Station, Showers, Laundry Nearby. Contact Grand Canyon National Park Lodges, Inc., Box 699, Grand Canyon, AZ 86023, Ph: 602-638-2401.

Lodges and Hotels

Bright Angel Lodge: On the Rim, Dining Room, Lounge, Fountain, Curios, Rooms & Cabins

El Tovar Hotel: Rooms, Double Occupancy, Dining Room, Lounge & Curios, On the Rim.

Kachina and Thunderbird Lodges: Standard Double Occupancy Rooms, On the Rim, Some View Rooms.

Maswik Lodge: Rooms and Cabins, Cafeteria, Lounge, Curios.

Yavapai Lodge: Double Occupancy Rooms, Cafeteria, Lounge and Curios. Closed in Winter.

Reservations for Above Lodges: Grand Canyon National Park Lodges, P.O. Box 699, Grand Canyon, AZ 86023, Ph: 602-638-2401

INNER CANYON

Reached by Foot or Mule Only. Overnite Permits Required.

Phantom Ranch Lodge: Cabins, Dormitories, Restaurant, Snack Bar. Advance Reservations Required—Contact: Grand Canyon National Park, Reservations Dept.
P.O. Box 699, Grand Canyon, AZ 86023
Ph: 602-638-2401

Campgrounds

Indian Gardens: 50 People at Designated Sites, 1 Group Site to 15 People

Bright Angel: 90 People at Designated Sites, 2 Group Sites to 31 People

Cottonwood: 40 People at Designated Sites, 1 Group Site to 13 People May Through October. Cottonwood has No Drinking Water or Ranger on Duty During Winter Months.

Each of the Above Campgrounds has Toilets, Drinking Water, Picnic Tables, Emergency Phones and Ranger Stations.

In addition to the above developed campgrounds, the Inner Canyon of Grand Canyon has been zoned as Threshold, Primitive and Undeveloped. Threshold Campsites Offer Minimal Facilities and are for the Experienced Canyon Hiker Only. Primitive and Undeveloped Zones are only for the Highly Experienced Canyon Hiker and are not recommended for use during the Summer Months due to Lack of Water and Extreme Heat.

For Permits and Reservations Write to: Backcountry Reservations Office, P.O. Box 129, Grand Canyon, AZ 86023

The Backcountry Information Line is Open from 11:00 a.m. to 5 p.m. Monday-Friday 602-638-2474. **No Reservations Accepted**

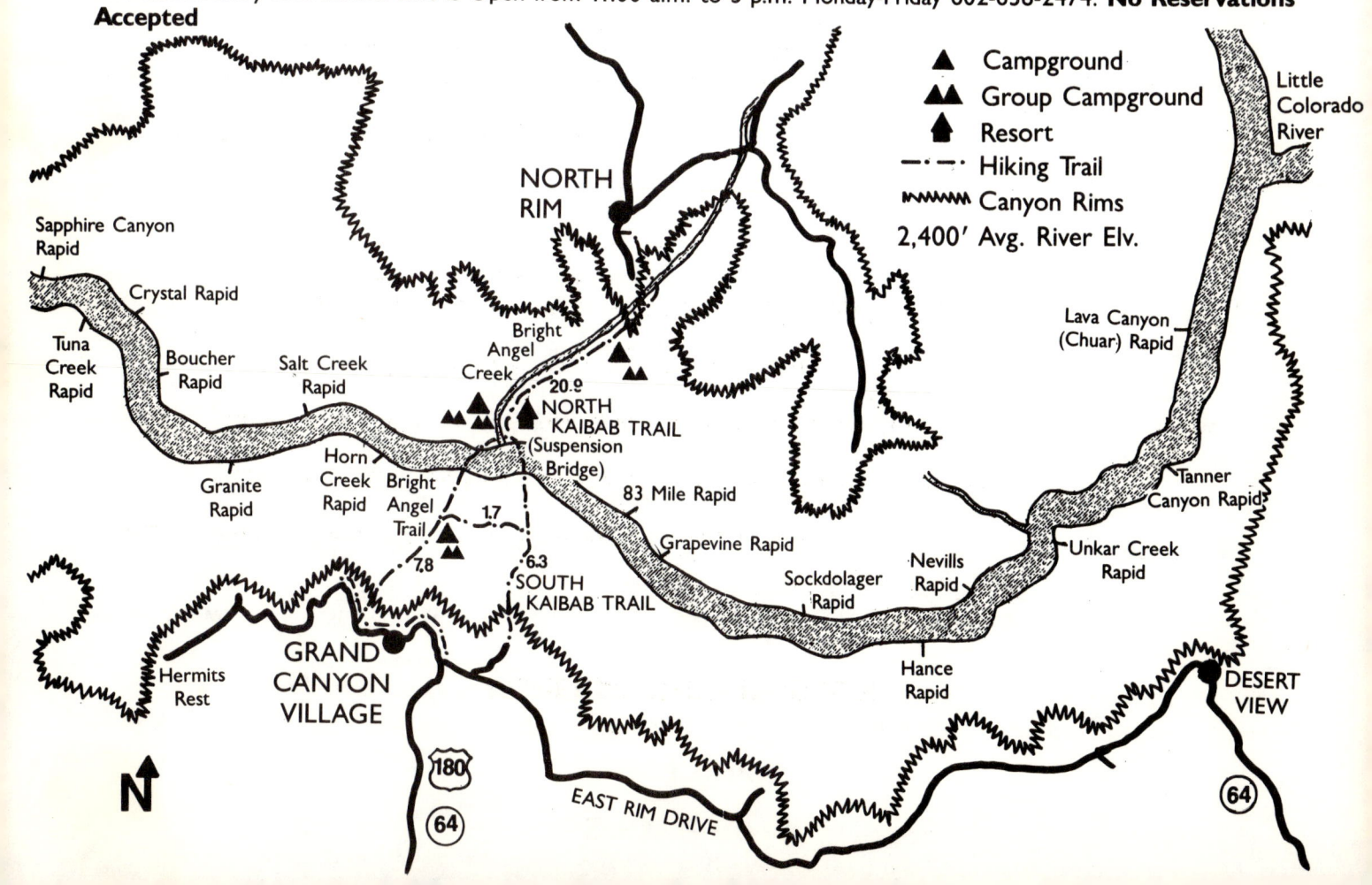

GENERAL INFORMATION

Grand Canyon Visitor's Center — hours vary according to season.

    Information Desk: General Information about the Park, Maps and Brochures

    Exhibit Hall: Exhibits on the Human and Natural History of the Grand Canyon

    Grand Canyon Orientation Program: Narrated Slide Shows

    Grand Canyon Natural History Bookstore: Publications and Maps on the Grand Canyon for Sale; scenic walks and informative lectures.

Yavapai Museum — Hours vary according to season.

    Exhibits on the Geological History of the Grand Canyon — a Geological Time Clock, Rock Samples and Explanations of the Formations of the Canyon and scenic walks.

For a Recorded Announcement of Daily Activities: Ph: 602-638-9304

Due to the Extensive Facilities in this Area
the Following is a List of Companies Operating Various Tours of the Grand Canyon

FLIGHTS

Plane

Air Grand Canyon P.O. Box 3028 Grand Canyon, AZ 86023 Ph: 602-638-2618	The Grand Tour	90 Minutes
	The Budget Tour	20 Minutes
	Grand Canyon as The Eagle Sees It	45 minutes

Grand Canyon Airlines P.O. Box 3038 Grand Canyon, AZ 86023 Ph: 602-638-2407 Outside Arizona: 1-800-528-2413	Canyon Highlights	45 Minutes
	Monument Valley/ Rainbow Bridge	3 Hours
	Transcanyon— Air Service Between South Rim & North Rim Eliminates a 5 hour Drive	20 Minutes— One Way

Williams Grand Canyon Flying Service Williams Municipal Airport 5000 N. Airport Road Williams, AZ 86046 Ph: 602-635-4711	The Grand Canyon Flight	1 Hour
	Havasupai Flight	1 Hour, 20 Min.
	Desert View Flight	1 Hour, 50 Min.

Helicopter

Grand Canyon Helicopters P.O. Box 455 Grand Canyon, AZ 86023 Ph: 602-638-2419 Outside Arizona: 1-800-528-2418	North Canyon Flight	30 Minutes
	Bright Angel Flight	20 Minutes
	Little Colorado	45 Minutes
Havasupai Flight	60 Minutes	

Madison Aviation Helicopters P.O. Box 536 Grand Canyon, AZ 86023 Ph: 602-638-2688	Canyon Flight	50 Miles
	Havasupai Flight	110 Miles

Helicopters (cont.)

Papillon Helicopters
Grand Canyon Airport
Grand Canyon, AZ 86023
Ph: 602-638-2431
Outside Arizona: 1-800-367-7095

The Scout Flight — Scenic Tour of Canyon's Central Portion, North Rim and Kaibab Plateau	30 Minutes
The Ranger — Scenic Tour of Havasupai Indian Reservation	60 Minutes
The Explorer — Complete Tour Including the Little Colorado River Gorge and 30 Minute Remote Landing	90 Minutes

---

Horseback and Mule Tours

Havasupai Canyon

Havasupai Tourist Enterprise
Supai, AZ 86435
Ph: 602-448-2121

Saddle & Pack Horses	
Hualapai Hilltop to Supai Village	$60 Round Trip
Hualapai Hilltop to Falls & Campgrounds	$80 Round Trip
Supai Village to Falls	$20 Round Trip

Mule Trips

Grand Canyon National
Park Lodges — Reservations
P.O. Box 699
Grand Canyon, AZ 86023
Ph: 602-638-2401

1 Day to Plateau Point:	12 Miles, 7 Hours
Overnight to Phantom Ranch: 4½ Hrs. Up	5½ Hrs. Down

Moqui Lodge

Apache Stables
P.O. Box 369
Grand Canyon, AZ 86023
Ph: 602-638-2424

East Rim Ride	4 Hours
Wagon or Horseback Ride to Campfire	1 Hour
Cowboy Breakfast Ride	

Bus Tours

Grand Canyon National Park Lodges — The Fred Harvey Transportation Company
South Rim Reservations
P.O. Box 699
Grand Canyon, AZ 86023
Ph: 602-638-2401

Full Day Indianlands Tour to Navajo-Hopi Reservation

Scenic Tour Along the South Rim Including the Watchtower, Hermit's Rest and Yavapai Museum

————————————

GRAND CANYON
LOWER CANYON

Leaving the concentration of facilities behind, the Colorado River resumes its isolated journey through the Grand Canyon. While the whitewater boater, awaits the fury of Vulcan Rapid (also called Lava Falls), Class 8 to 10, a drive to remote Toroweap Point provides a spectacular vista of the River 2,802 feet below, as well as the Western reaches of the Canyon. A trip to Havasu Falls is a somewhat difficult, but rewarding experience. You must park your car and hike ten miles to the Havasaupai Village of Supai, 2,300 feet below. Helicopters also fly in from Grand Canyon, and the Havasaupai provide pack animals. Reservations are required to enter Havasu. A few commercial outfitters fly into Whitmore Canyon and depart on a leisurely tour through the remote lower canyons to Lake Mead.

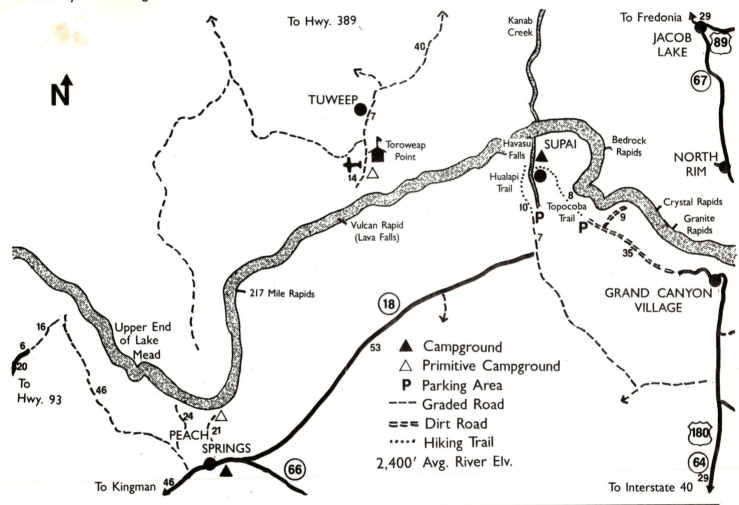

INFORMATION: Superintendent, Grand Canyon National Park, Box 129, Grand Canyon, AZ 86023
Ph: 602-638-7888

CAMPGROUNDS & RESORTS

Havasupai Indian Reservation—Havasupai Tourist Enterprise, Supai, AZ 86435 Ph: 602-448-2121. 300 Tent Sites, Drinking Water, Restrooms, Pack Horses, Pack In/Pack Out

Grand Canyons Caverns: Peach Springs, AZ 86434, Ph: 602-422-3223, 119 Tent/R.V. Sites, 19 Full Hookups, Disposal Station, Grocery Store, Gas, Riding Trails, Local Tours

RECREATION

Whitewater Boating: See Previous Pages

Commercial River & Air Tours

Fishing—River & Streams: Rainbow & Native Trout

Hunting—Kaibab National Forest: Deer

Hiking & Backpacking: See Previous Pages

Picnicking

Nature & Geological Study

Indian Village & Artifacts

OTHER

National Park Service: Toroweap Point: 10 Tent/R.V. Sites, No Water, Primitive

HYPOTHERMIA

If a person falls into the River, hypothermia should be watched for. The thermal conductivity of water is 200 times greater than that of air. Emersion in cold water rapidly lowers the body temperature. When a person has been in the water for an extended period, muscular strength and co-ordination is rapidly diminished after 5 minutes. After 10 to 15 minutes, a person is totally unable to help himself. The symptoms of hypothermia are muscular unco-ordination, uncontrollable shivering and mental impairment.

It is extremely important to get the victim out of the water as soon as possible. Remove all wet clothing and dress him with warm, dry clothing. A hypothermic person has lost the ability to rewarm himself and must be supplied an external heat source. Placing a second person into a sleeping bag with him provides direct body-to-body contact and is one of the best sources of this external heat. When the victim is able to eat, feed him high energy foods which restores the vital energy needed to maintain body temperature.

DRINKING WATER

Do not drink untreated water! Giardiasis, an intestinal disorder, is reaching epidemic proportions. Giardiasis lambia is a small microscopic organism found in many lakes, streams and rivers. It can cause extreme discomfort and must be treated by a physician.

Where drinking water is not available, bring your own or be sure you treat it. While on the River, refill your water containers from the center of the River, never from tributaries or near the shore where contamination is at its highest. Always add the prescribed amount of a clorinating agent (8 to 16 drops per gallon of water, depending on whether it is clear, cloudy or turbid), mix thoroughly and let stand for at least thirty (30) minutes. Another method of treatment is to boil the water at least one (1) minute at sea level and for five (5) minutes at higher elevations.

Giardia is easily transmitted between humans and animals. Always use good sanitary precautions. All feces, human and animal, should be buried at least one hundred feet away from natural waters. Protect those who follow by keeping our rivers, lakes and streams free from contamination.

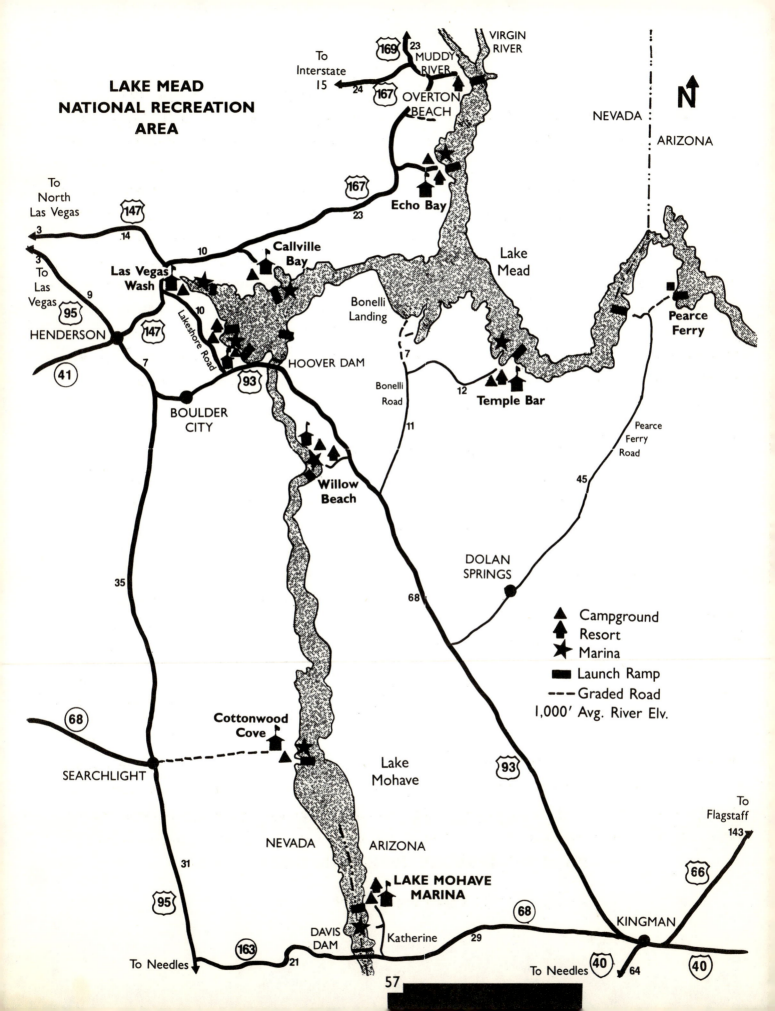

LAKE MEAD
NATIONAL RECREATION
AREA

To Interstate 15

169 23 MUDDY RIVER
VIRGIN RIVER
24 167 OVERTON BEACH
NEVADA
ARIZONA
N

To North Las Vegas
147
14
167
23
Echo Bay
Lake Mead
3
10
Callville Bay
Las Vegas Wash
3 To Las Vegas
9
95
Bonelli Landing
Pearce Ferry
HENDERSON
147
Lakeshore Road
10
7
Bonelli Road
7
Temple Bar
41
7
HOOVER DAM
12
93
BOULDER CITY
11
Pearce Ferry Road
45
Willow Beach
35
DOLAN SPRINGS
68
Campground
Resort
Marina
Launch Ramp
Graded Road
1,000' Avg. River Elv.
68
Cottonwood Cove
SEARCHLIGHT
Lake Mohave
93
To Flagstaff
143
31
NEVADA
ARIZONA
66
95
LAKE MOHAVE MARINA
163
DAVIS DAM
Katherine
68
KINGMAN
To Needles
21
29
To Needles
40
64
40

57

LAKE MEAD NATIONAL RECREATION AREA

LAKE MEAD — LAKE MOHAVE

For millions of years, the Colorado River tenaciously cut its way through this area of rugged mountains and desert. Free to carve at will the awesome canyons on its way to the Gulf of California, this tempermental River held sway over the environment. For centuries, man's efforts to exploit its resources were thwarted by floods and low water. The River was always in command. In 1935 man made a major stride in assuming control.

The first Dam on the Colorado was dedicated on September 30, 1935. After years of political battles and technical problems, Hoover Dam became a reality. Many felt it couldn't and shouldn't be done. Reputable engineers said it was technically impossible, but a group of dedicated engineers and construction men from the Bureau of Reclamation and the Six Companies, Incorporated, proved it could. This greatest of engineering feats of that time resulted in a Dam 727 feet high anchored on a concrete base 660 feet thick between the volcanic walls of Black Canyon. Behind the Dam, Lake Mead stretches 110 miles even reaching the lower part of Grand Canyon. There is a total surface area of 157,900 acres with a maximum depth of 585 feet at full capacity.

Davis Dam was completed in 1953 in compliance with the Mexican Water Treaty of 1944. The Dam is located 67 miles downstream from Hoover Dam in Pyramid Canyon just 10 miles north of the point where Arizona, Nevada and California meet. The resulting narrow Lake Mohave is confined by the sheer walls of Pyramid, Painted, Eldorado and Black Canyons. The widest point on this 44 square mile Lake is four miles, thereby retaining a lot of the characteristics of the River.

Today, the mighty Colorado provides hydroelectric power and water benefiting millions of people. Over 2 years of River flow (32 million acre feet) rest in Lake Mead. Man now has reasonable control over the floods and draught created by the demands of the River. Hoover and Davis Dams have provided an abundance of benefits, not the least of which is the recreation wonderland of Lake Mead and Lake Mohave.

LAKE MEAD NATIONAL RECREATION AREA

LAKE MEAD — LAKE MOHAVE

(Continued)

Lake Mead and Lake Mohave comprise the Lake Mead National Recreation Area under the administration of the National Park Service. The Park Service has installed hazard buoys, navigation lights and mile post markers. Winds can be a hazard so always check the weather forecast and keep your safety equipment available and in good order. Be prepared for emergency. The Lakes' rules are enforced by the Nevada and Arizona Departments of Fish and Game, the Park Service, and the U.S. Coast Guard.

The Lake Mead National Recreation Area is a boater's paradise. While the surrounding desert lands lure the hunter and naturalist, the vast majority of the 6 million yearly visitors are attracted to the waters of Lake Mead and Lake Mohave. Sailing is excellent. There is virtually unlimited boating from waterskiing to houseboating. Fishing is a prime attraction, as is swimming or just plain relaxing in this warm desert environment. Each Lake shares this year round recreational abundance, yet each has its own character.

INFORMATION: Lake Mead National Recreation Area, 601 Nevada Highway, Boulder City, NV 89005, Ph: 702-293-4041

CAMPGROUNDS & RESORTS	RECREATION	OTHER
For Developed Campsites and Resorts, See Following Pages Undeveloped Backcountry Campsites are Shown on Following Maps Boat Camping is Open Throughout Recreation Area Except for Designated Areas. 90-Day Limit	Boating: Open to All Boating, Full Service Marinas & Rentals. 24-Hour Emergency Phone: 702-293-4041 Fishing: Largemouth & Striped Bass, Rainbow & Cutthroat Trout, Channel Catfish, Crappie & Bluegill, Silver Salmon & Green Sunfish Hunting: Deer, Big Horn Sheep Swimming & Scuba Diving Backcountry (No Trails) Hiking & Horseback Riding & Nature Study	Full Facilities & Casinos in Las Vegas and Other Nearby Cities Central Reservations for Playmate Resorts-Marinas 730 S. Cypress La Habra, CA 90631 Ph: 213-691-2235 or 714-871-1476

LAKE MEAD NATIONAL RECREATION AREA
TEMPLE BASIN

Rafters, spent and satiated from their run through the Grand Canyon, leisurely pass The Temple at the end of their journey. Greeting them are boaters enjoying an adventure of another choice. Power boaters during periods of high water journey into the lower reaches of the Grand Canyon carefully avoiding the hazards of floating debris and sand bars. Waterskiers speeding past, relish the vast open expanse of clear blue water. Fishermen can boat stripers, largemouth and rainbows. This relatively remote section of Lake Mead not only offers an abundance of water-oriented opportunities, but also the lure of the desert with it springtime wild flower display, animals, minerals and harsh environment. Those entering this back-country are cautioned to stay on the designated roads and camp only in designated areas.

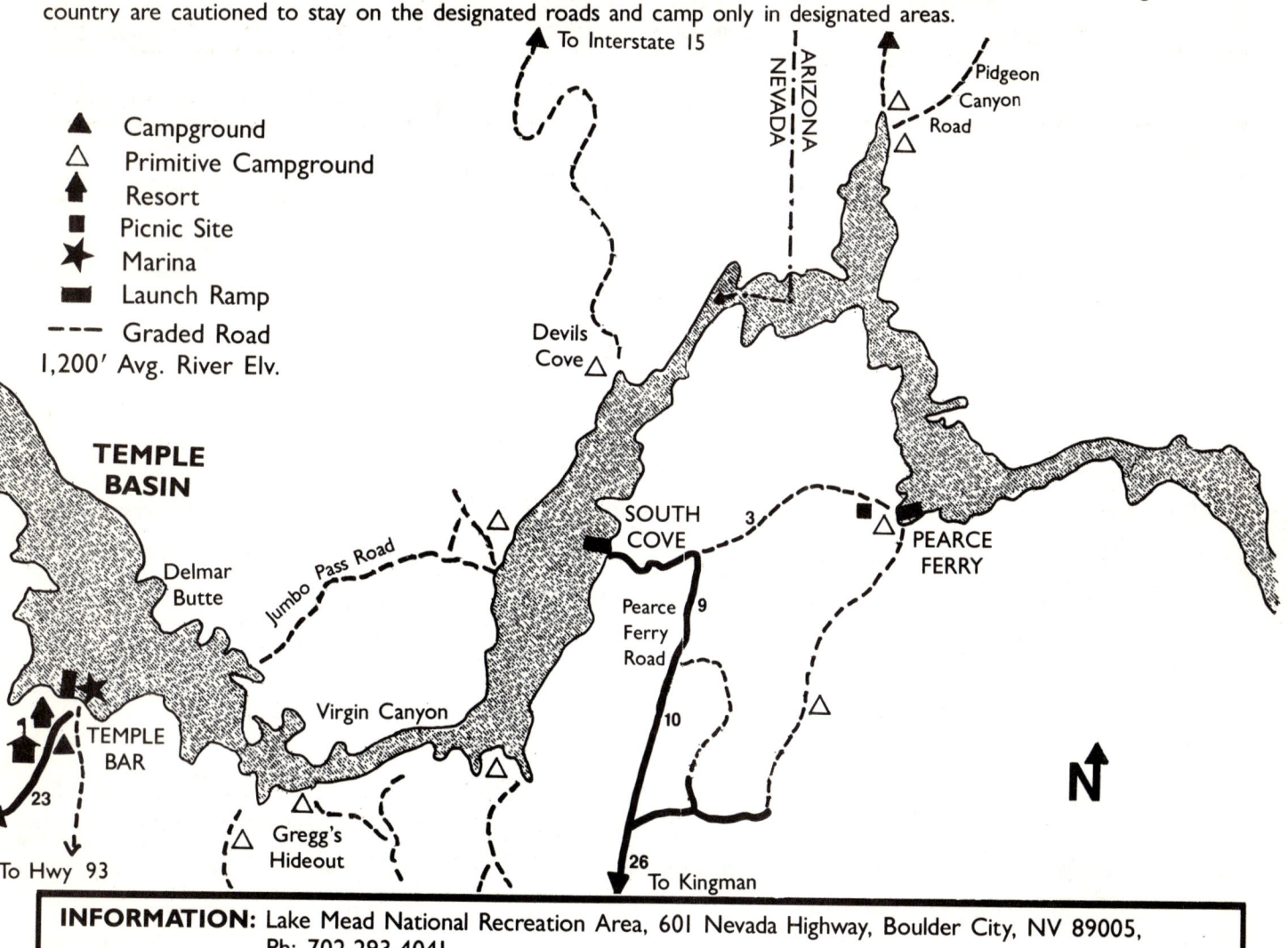

CAMPGROUNDS & RESORTS	RECREATION	OTHER
Temple Bar Campground—153 Tent/R.V. Sites, No Hookups, Disposal Station, Recreation Programs **Temple Bar Resort**—Temple Bar, AZ 86443, Ph: 602-767-3400. R.V. Park, Full Hookups, Motel, Housekeeping Cabins, Store, Restaurant & Lounge	Boating: Open to All Boating, Full Service Marina, Slips & Storage Rentals: Waterski, Fishing & Patio Boats, Skis & Fishing Gear See Previous Graph for Other Information	Primitive Campsite as Noted on Map

INFORMATION: Lake Mead National Recreation Area, 601 Nevada Highway, Boulder City, NV 89005, Ph: 702-293-4041

LAKE MEAD NATIONAL RECREATION AREA
VIRGIN BASIN

The wide open area of water in Virgin Basin provides ample room for sailing, boating and waterskiing. While normal winds on Lake Mead are in the 15-20 mile per hour range, weather conditions can change with little or no warning, bringing on high winds and waves of 6 to 8 feet. Caution is advised. Weather forecasts are made twice daily by the Park Service and are available at Ranger Stations, concessions and launch ramps. Facilities at Virgin Basin are limited to undeveloped campgrounds and launch ramps at Bonelli Landing and Detrital Wash. Houseboaters seek out one of the numerous coves and inlets to enjoy a quiet retreat to fish, swim or relax.

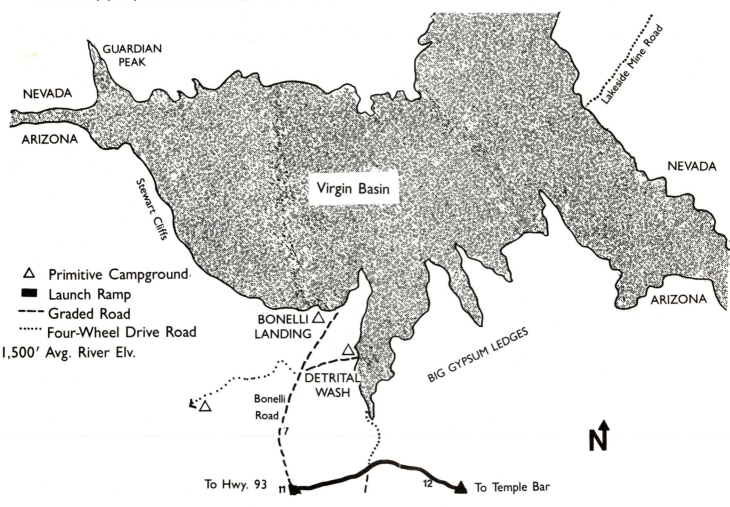

INFORMATION: Lake Mead National Recreation Area, 601 Nevada Highway, Boulder City, NV 89005, Ph: 702-293-4041		
CAMPGROUNDS & RESORTS	**RECREATION**	**OTHER**
Undeveloped Campsites as Shown on Map	See Previous Pages	

LAKE MEAD NATIONAL RECREATION AREA
OVERTON ARM

The Overton Arm of Lake Mead reaches 65 miles north to the mouth of the Virgin River. This historical and scenic area was once the home of ancient Pueblos. Many of their ruins are now under the water of Lake Mead. A reconstructed Pueblo Village along with a display of their artifacts can be seen at the Lost City Museum in Overton. Just west of Overton, the Valley of Fire State Park provides an interesting view of Indian petroglyphs and unusual geological formations. This section of the Lake offers excellent accommodations and marine facilities. There are good water sports and birding opportunities.

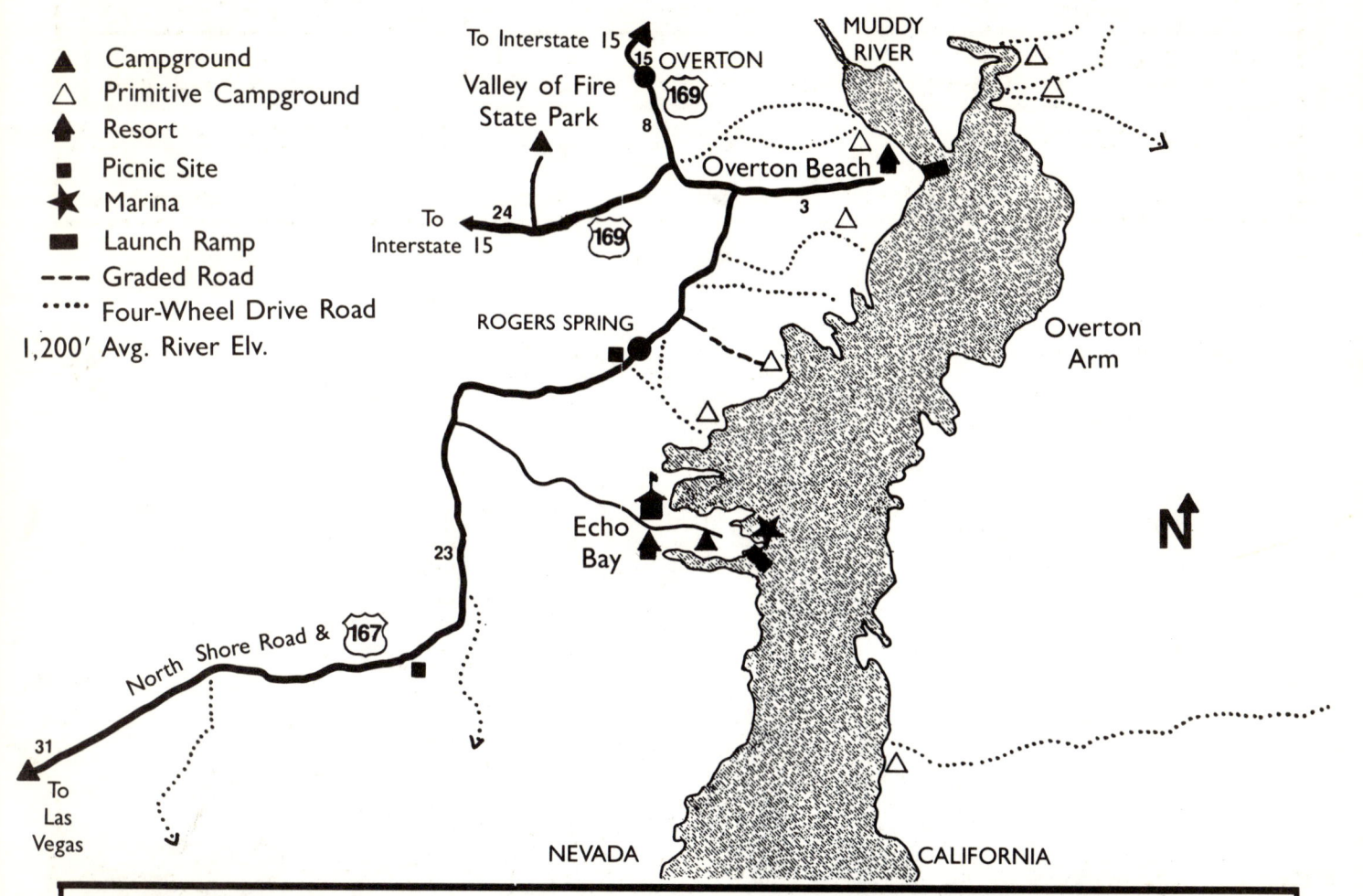

▲	Campground
△	Primitive Campground
♠	Resort
■	Picnic Site
★	Marina
◼	Launch Ramp
---	Graded Road
····	Four-Wheel Drive Road

1,200' Avg. River Elv.

INFORMATION: Lake Mead National Recreation Area, 601 Nevada Highway, Boulder City, NV 89005, Ph: 702-293-4041

CAMPGROUNDS & RESORTS

Echo Bay Campground—80 Tent/R.V. Sites, Water, Tables (30 day Limit). National Park Service.

Valley of Fire State Park—Box 515, Overton, NV 89040. Ph: 702-397-2088
2 Campgrounds, 47 Sites, Tables, Water, Grills & Restrooms, Disposal Station

Undeveloped Campsites as Shown on Map

RECREATION

Overton Beach Resort—Paved Launch Ramp, Boat & Motor Rentals, Guide Service

Echo Bay Marina—Full Service Marina, Moorings, Fuel Dock, Slips, Paved Launch Ramp. Rentals: Fishing and Waterski Boats, Houseboats

OTHER

Echo Bay Resort—
Overton, NV 89040,
Ph: 702-394-4066
R.V. Park, Full Hookups, Large Grocery Store, Modern Hotel, Dining Room and Cocktail Lounge

Overton Beach Resort
Overton, NV 89040,
Ph: 702-394-4040
R.V. Park, Cafe, Bait & Tackle Swimming Beach

LAKE MEAD NATIONAL RECREATION AREA
BOULDER BASIN

Boulder Basin is the busiest section of Lake Mead. Numerous facilities, campgrounds and marinas await the visitor to this most western part of the Lake. The Bureau of Reclamation conducts daily guided tours of Hoover Dam. Over 600,000 visitors a year enjoy the tour and exhibits of this engineering marvel. Lake Mead Yacht Tours provide a scenic tour of the Lake. The excursion cruise departs from the Lake Mead Marina at 10:30 a.m., 12:00 Noon, 1:30 p.m. and 3:00 p.m. Numerous other attractions are found in Henderson and Boulder City. Nearby Las Vegas provides an overwhelming diversion. In spite of this competition, the waters of Lake Mead are the primary attraction with its wealth of recreational opportunities.

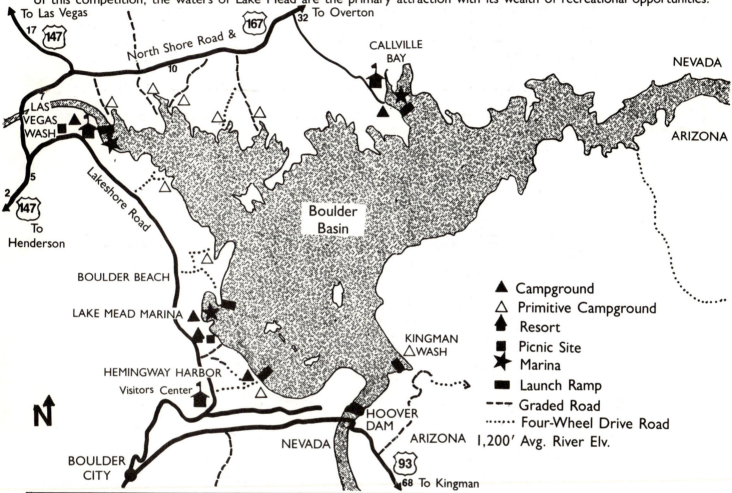

INFORMATION: Lake Mead National Recreation Area, 601 Nevada Highway, Boulder City, NV 89005, Ph: 702-293-4041

CAMPGROUNDS & RESORTS	RECREATION	OTHER
Boulder Beach—154 Tent/R.V. Sites, Disposal Station, Hot Showers, Restaurant & Cocktail Lounge, General Store Nearby	**Lake Mead Resort & Marina**—322 Lakeshore Rd., Boulder City, NV 89005 Ph: 702-293-2074 Lodge, Floating Restaurant & Lounge, Full Service Marina, Slips & Launch Ramp, General Store. Rentals: Fishing & Water-ski Boats, Houseboats, Yacht Tours	**Lakeshore Trailer Village**—268 Lakeshore Rd., Boulder City, NV 89005 Ph: 702-293-2540 75 R.V. Sites, Full Hookups, Hot Showers, Laundry, Store, Snack Bar
Callville Bay—80 Tent/R.V. Sites, Disposal Station, General Store Nearby		**Callville Bay**—Star Rt. 10, Box 100, Las Vegas, NV 89124 Ph: 702-565-8958. 150 Site R.V. Park, Full Hookups, General Store, Restaurant & Lounge, Full Service Marina. Rentals: Sail, Fishing, Power Boats, Houseboats
Hemingway—184 Tent/R.V. Sites, General Store Nearby		
Las Vegas Wash: 89 Tent/R.V. Sites, Disposal Station		

LAKE MOHAVE

This intimate section of Lake Mohave is confined between the walls of narrow canyons retaining many of the characteristics of the original River. Even rapids are found at low water levels at the upper end of Black Canyon. Water oriented birds and animals are seen on the shoreline. Numerous coves and inlets attract the boater who might spot a desert Bighorn on the cliffs above. The cold, clear water flowing from Hoover Dam has produced record sized rainbows although the principal game fish is largemouth bass. The Willow Beach Hatchery plants trout the year around. There are good support facilities at Willow Beach and Cottonwood Cove.

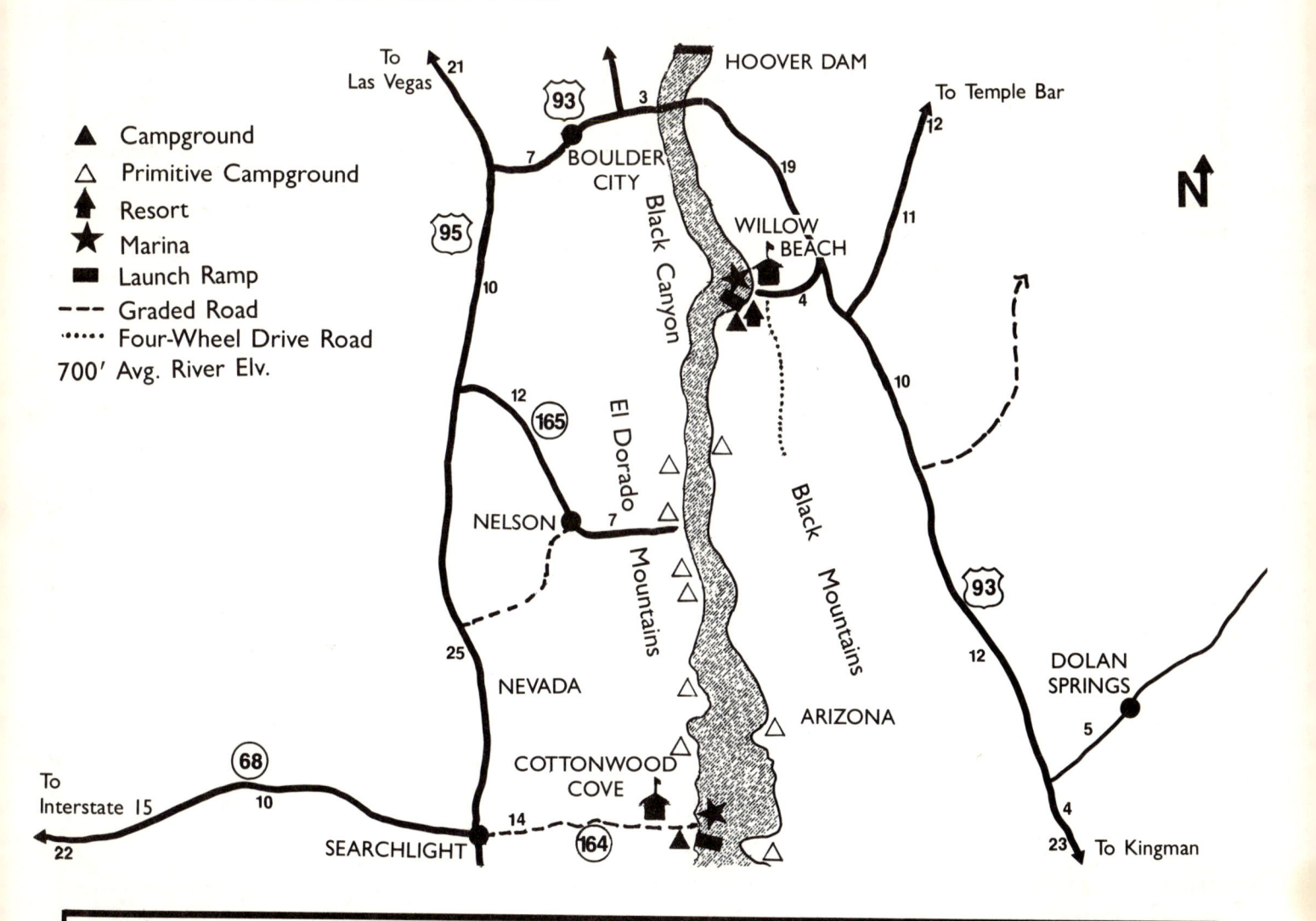

▲	Campground
△	Primitive Campground
🏠	Resort
★	Marina
■	Launch Ramp
- - -	Graded Road
····	Four-Wheel Drive Road

700' Avg. River Elv.

INFORMATION: Lake Mead National Recreation Area, 601 Nevada Highway, Boulder City, NV 89005
Ph: 702-293-4041

CAMPGROUNDS & RESORTS	RECREATION	OTHER
Willow Beach Resort—P.O. Box 187, Boulder City, NV 89005 Ph: 702-767-3311 15 R.V. Sites, Full Hookups, Motel, Restaurant, Store **Cottonwood Cove**—P.O. Box 1000, Cottonwood Cove, NV 89046 Ph: 702-297-1464 145 Tent/R.V. Sites, 75 Full Hookups, General Store, Restaurant, Motel	**Willow Beach**—Full Service Marina, Launch Ramp, Slip Rentals, Patio and Fishing Boat Rentals **Cottonwood Cove**—Full Service Marina, Slip Rentals, Launch Ramp, Houseboat, Fishing and Power Boat Rentals	Facilities in Boulder Creek and Searchlight

LAKE MOHAVE — DAVIS DAM

Leaving the narrow canyons above Cottonwood Valley, Lake Mohave expands to its widest point of four miles. The 237 miles of everchanging shoreline of sandy beaches, rock cliffs and secluded coves offer a beautiful contrast to the clear blue water of the Lake. Relatively limited road access leaves the boater almost exclusive domain to cruise, waterski or just relax on a houseboat. While fishing for trout and largemouth bass is the most popular, stripers, channel cats and crappie await the interested angler. The surrounding area offers good hunting for cottontail and jack rabbits. A special drawing provides a rare opportunity to bag a Bighorn Sheep. Excellent accommodations and marine facilities await the visitor to this water wonderland.

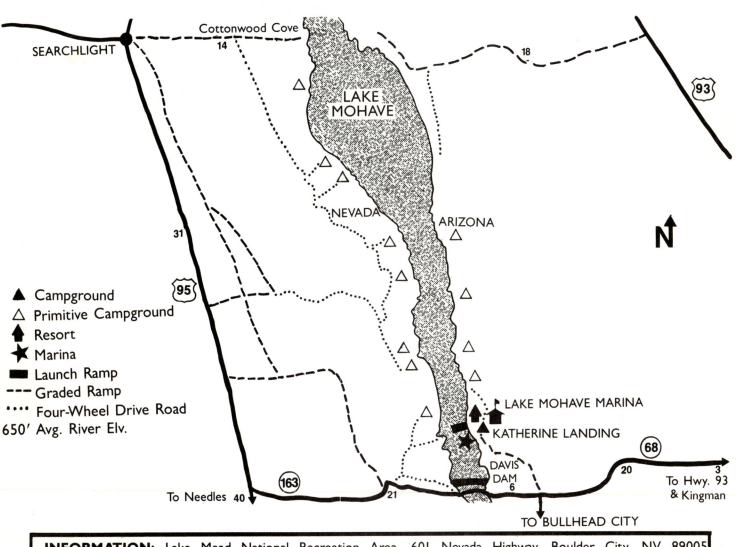

▲ Campground
△ Primitive Campground
⬆ Resort
★ Marina
◼ Launch Ramp
--- Graded Ramp
•••• Four-Wheel Drive Road
650' Avg. River Elv.

INFORMATION: Lake Mead National Recreation Area, 601 Nevada Highway, Boulder City, NV 89005
Ph: 702-293-4041

CAMPGROUNDS & RESORTS	RECREATION	OTHER
Katherine Campground—173 Tent/R.V. Sites, General Store, Restaurant (National Park Service) **Lake Mohave Resort**—Katherine Landing, Bullhead City, AZ 86430. R.V. Park, Motel, Restaurant & Lounge, Store	**Lake Mohave Marina**—Full Service Marina, Slips and Launch Ramp, Houseboat, Waterski, Fishing and Patio Boat Rentals	Full Facilities in Bullhead City

DAVIS DAM TO FORT MOHAVE

The Colorado River again returns to its natural state after resting in the Lakes above. The cold, clear water flowing out of Davis Dam provides an ideal habitat for large rainbow trout. This section of the River holds the world's record for landlocked striped bass of 59¾ pounds. If that's not enough, the angler will also find good catfishing. Boating and waterskiing are very popular. Sportsmen flock to this area during the winter months. Laughlin, Nevada, just across the River by free ferry, is one of the fastest growing gaming centers in the United States. It offers a number of good hotels, casinos and live entertainment.

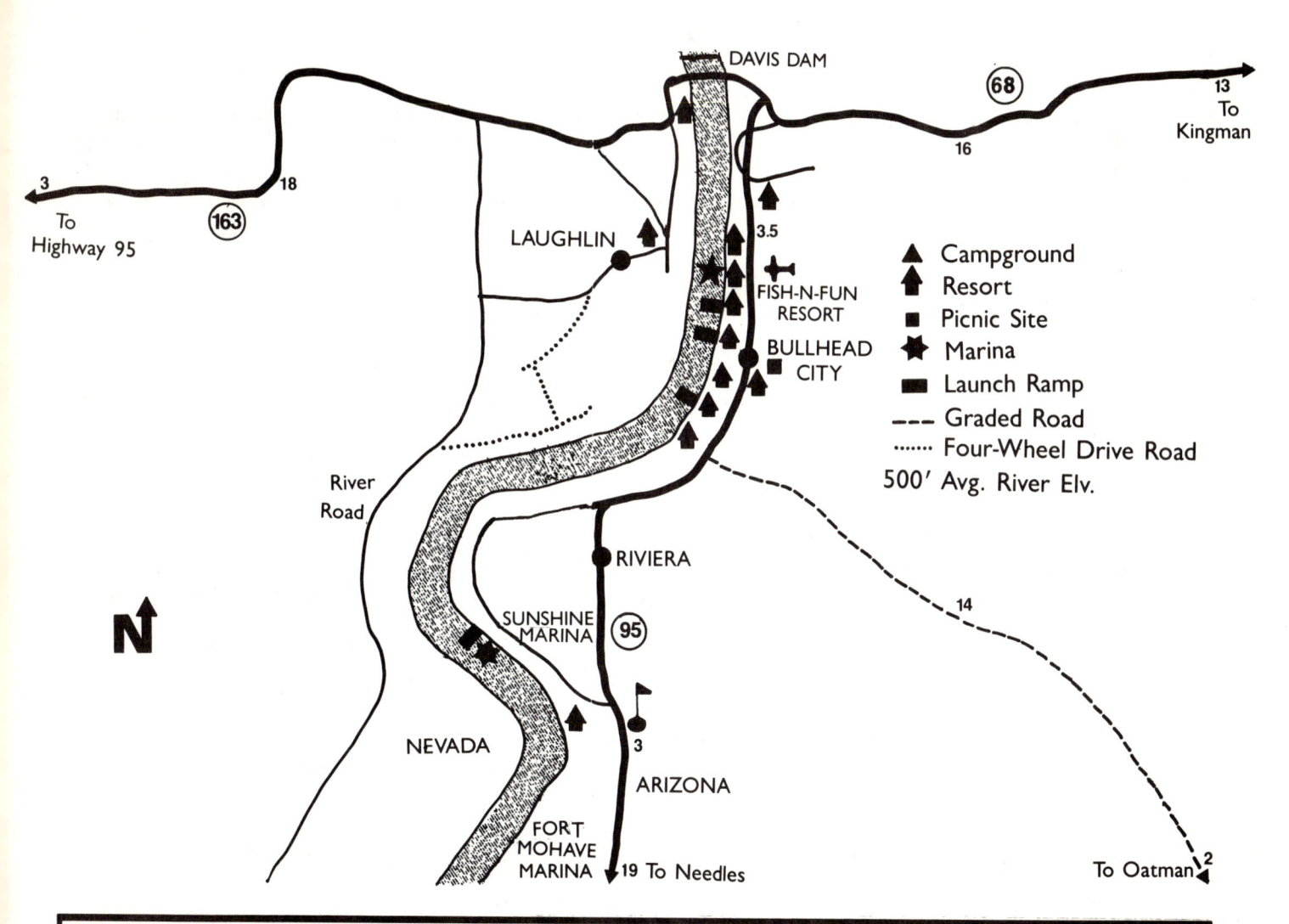

INFORMATION: Bullhead City Chamber of Commerce, P.O. Box 66, Bullhead City, AZ 86430, Ph: 602-754-3891

CAMPGROUNDS & RESORTS	RECREATION	OTHER
See Following Page	Boating: Open to All Boating, Canoe and Float Trips, Drag Races Fishing: Striped Bass, Channel Catfish and Rainbow Trout Hunting: Cottontail Rabbits, Quail, Dove and Waterfowl Picnicking Swimming Rockhounding Golf	Full Facilities in Bullhead City and Laughlin For Casino and Resort Information in Laughlin: Ph: 800-227-0629—Arizona 800-227-5245—Outside Arizona

DAVIS DAM TO FORT MOHAVE

(Continued)

Arizona Resorts

B Bar B R.V. Resort, 4750 Highway 95, Bullhead City, AZ 86430, Ph: 602-763-4664.
  87 R.V. Sites, Full Hookups, Separate Tent Area, Flush Toilets, Hot Showers, Laundry and Patios.

Bullhead R.V. Resort, 1000 Highway 95, Bullhead City, AZ 86430, Ph: 602-763-8353.
  156 R.V. Sites, Full Hookups, Disposal Station, Flush Toilets, General Store, LP Gas, Swimming Pool, Adults.

Carefree Resort, 350 Lee, Bullhead City, AZ 86430, Ph: 602-754-3438.
  52 R.V. Sites, Full Hookups, Flush Toilets, Showers, Laundry, Patios, Swimming Pool, Whirlpool, Adults.

El Rio Motel and Trailer Park, 1063 Highway 95, P.O. Box 102, Bullhead City, AZ 86430, Ph: 602-763-4385.
  20 R.V. Sites, Full Hookups, Flush Toilets, Motel, Paved Ramp and Boat Dock.

Fish-N-Fun Trailer Resort, P.O. Box 587, Bullhead City, AZ 86430, Ph: 602-754-3999.
  26 R.V. Sites, Full Hookups, Flush Toilets, Hot Showers, Laundry, Boat Dock, Ramp.

Hardyville Manor Trailer Park, 1753 Georgia Lane, Riviera, AZ 86442, Ph: 602-754-3994.
  17 R.V. Sites, Flush Toilets, Hot Showers, Laundry, Fishing Dock.

KOA, Highway 95 & Merrill Ave., Bullhead City, AZ 86430, Ph: 602-763-2179.
  89 Full Hookups, 18 Water and Electric Hookups, Separate Tent Area, Flush Toilets, Hot Showers, Disposal Station, Laundry, General Store, L.P. Gas, Picnic Tables, Swimming and Wading Pool.

Ridgeview Park R.V. Resort, 2751 Locust Blvd., Bullhead City, AZ 86430, Ph: 602-754-2595.
  300 R.V. Sites, Full Hookups, Cable T.V., Swimming Pool and Spa.

Sunset Landing, 2775 Camino de Rio, Riviera, AZ 86442, Ph: 602-758-3133.
  6 Tent, 12 R.V. Sites, Full Hookups, Flush Toilets, Hot Showers.

Nevada Resorts

Sportsman's Park, Clark County Parks and Recreation, Ph: 702-386-4384.
  61 Tent/R.V. Sites, Flush Toilets, Cold Showers, Boat Ramp, First Come—First Served.

Riverside Resort, P.O. Box 500, Laughlin, NV 89046, Ph: 702-298-2535.
  300 Space R.V. Park, Security Patrol, Full Hookups, Laundry, Showers, Swimming Pool, Restaurant, Hotel and Casino.

Marinas

Fish-N-Fun Resort, P.O. Box 587, Bullhead City, AZ 86430, Ph: 602-754, 3999.
  52 R.V. Sites, Full Hookups, Hot Showers, Laundry, Flush Toilets, Fishing Dock, Recreation Hall, Boat Ramp, No Pets.

Sunshine Marina, 2170 Kaibab Avenue, Riviera, AZ 86442, Ph: 602-758-6322.
  Berths, Fuel Dock, Launch Ramp, Restaurant & Cocktail Lounge.

Public Ramps at end of 1st St. (Paved), 2nd St. (Gravel), 3rd St. (Dirt) and 4th St. (Paved), Bullhead City, AZ.

NEVADA

68

To Albuquerque

Kingman

40

144

Flagstaff

15

105

40

207

21

Lake Havasu City

73

CALIFORNIA

95

95

62

54

ARIZONA

141

Parker

SAN BERNARDINO

48

49

72

17

23

10

23

95

215

184

Blythe

95

10

PHOENIX

15

78

150

108

118

COLORADO RIVER

95

63

10

24

83

El Cajon

117

8

El Centro

SAN DIEGO

57

8

RIVER GILA

Yuma

To Tucson

193

Tijuana

Mexicali

181

25

116

73

5

San Luis

2

Ensenada

39

113

1

95

459

MEXICO

PACIFIC OCEAN

N

To San Felipe

GULF OF CALIFORNIA

To Santa Ana

FORT MOHAVE TO NEEDLES

This section of the Colorado River meanders through the fast developing Mohave Valley. Located in the heart of the Lower Colorado's vacationland, the warm, dry climate and desert scenery make this area a prime recreational attraction. The River is a boater's paradise with over 100 miles of sheltered coves and navigable waterway. Swimming and sunbathing are popular, and the fishing is excellent. Rockhounds and Western buffs enjoy nearby Oatman. This historic mining town, the site of many Western movies, has maintained many of its original buildings. Wild burros freely roam the streets. Needles has three fine marinas which are the site of many regattas, races and cruises.

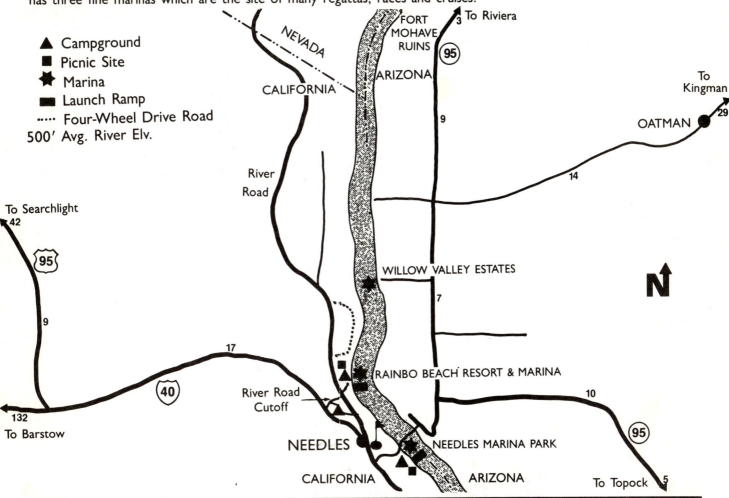

▲ Campground
■ Picnic Site
★ Marina
▬ Launch Ramp
···· Four-Wheel Drive Road
500' Avg. River Elv.

INFORMATION: Needles Chamber of Commerce, P.O. Box 705, Needles, CA 92363, Ph: 619-326-2050.

CAMPGROUNDS & RESORTS	RECREATION	OTHER
KOA Needles—5400 National Old Trails Highway, Needles, CA 92363, Ph: 619-326-4207. 55 R.V. Sites, Full Hookups, Disposal Station, Showers, Laundry, Propane, Groceries, Swimming Pool. **Needles Park Marina**—100 Marina Dr., Needles, CA 92363, Ph: 619-326-2197. 190 R.V. Sites, Full Hookups, Storage, Showers, Disposal Station, Laundry, Groceries, Supplies, Swimming Pool, Whirlpool, Boat Dock, Ramp & Fuel.	Open to All Boating: Canoe and Float Trips Fishing: Striped Bass, Channel Catfish and Rainbow Trout Hunting: Cottontail Rabbits, Quail, Dove & Waterfowl Picnicking Swimming Hiking & Nature Study Rockhounding Golf	**Rainbo Beach Marina**—Ph: 619-326-3101, 70 R.V. Sites, Full Hookups, Showers, Handicap Facilities, Laundry, Swimming Pool, Whirlpool, Picnic Tables and Patios, Boat Dock, Ramp and Fuel. Full Facilities in Needles.

TOPOCK MARSH AND TOPOCK GORGE

The recreational abundance of the Coloardo is enhanced in this area of the River by the Topock Marsh. This Marsh is a part of the Havasu National Wildlife Area, and is a sportsman's paradise. Although boaters are cautioned to be aware of underwater stumps and snags, many boaters and canoers enjoy this wetlands experience. Birds and wildlife are abundant. The fishing and frogging are excellent. The hunter will find cottontail and jackrabbits, quail, dove and waterfowl. Canoeing and floatboating are popular on the River from Davis Dam to Lake Havasu. A day trip through the 16 mile Topock Gorge provides many scenic rewards. Camping is restricted to designated campgrounds in the Refuge. There is no camping in Topock Gorge.

▲ Campground
■ Picnic Site
★ Marina
▬ Launch Ramp
--- Graded Road
····· Four-Wheel Drive Road
500' Avg. River Elv.

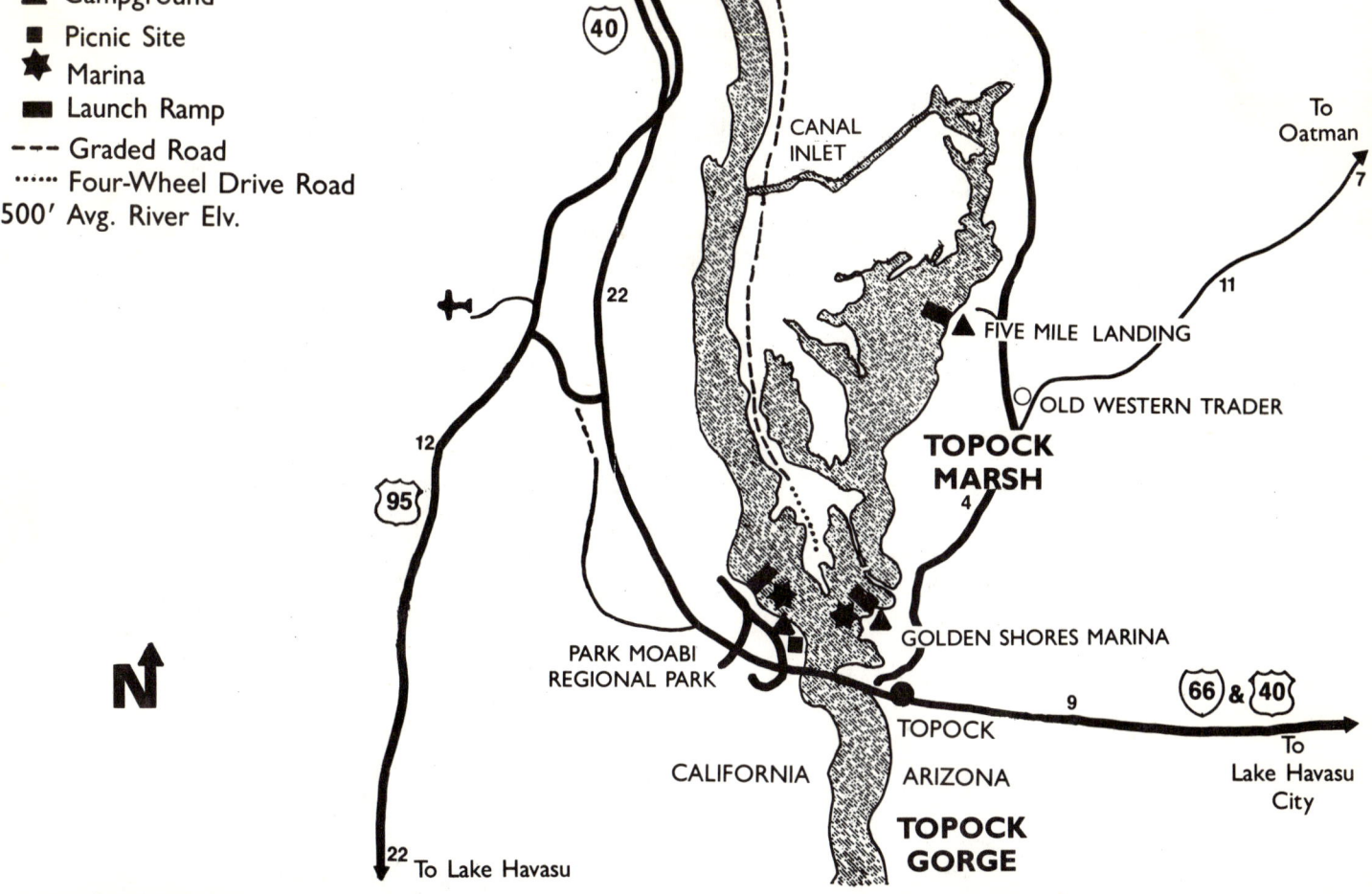

INFORMATION: Havasu National Wildlife Refuge, P.O. Box A, Needles, CA 92363, Ph: 619-326-3853

CAMPGROUNDS & RESORTS

Park Moabi Regional Park—Park Moabi Rd., Needles, CA 92363, Ph: 619-326-3831. 648 Tent/R.V. Sites, Full & Partial Hookups, Disposal Station, Hot Showers, Picnic Tables, Grills, Deli and General Store, Swim Beach, Boat Ramp, Gas Docks, Houseboat & Boat Rentals.

Golden Shores Marina—Ph: 602-768-2325. 40 R.V. Sites, Full Hookups, Store, Restaurant, Gas, Boat Fuel & Ramp.

RECREATION

Boating: Open to All Boating **except** Waterskiing in Topock Gorge & Marsh. Float Trips & Canoeing.
Fishing: Rainbow Trout, Striped & Largemouth Bass, Channel Catfish, Crappie, Bluegill & Bullfrogs.
Hunting: Cottontail & Jack Rabbits, Quail, Dove & Waterfowl
Swimming & Skin Diving
Picnicking & Hiking
Birding & Nature Study

OTHER

5-Mile Landing—P.O. Box 48, Topoc, AZ 86436, Ph: 602-768-2350. 30 Tent, 75 R.V. Sites, Full Hookups, Showers, Laundry, Grocery Store, Gas, Boat Ramp, Fuel Dock. Rentals: Row, Canoe, Pedal.

Old Western Trader—Box 350, Topock, AZ 86436. Canoe Rentals, Delivery & Car Shuttle, Bait & Tackle, Laundry, Store.

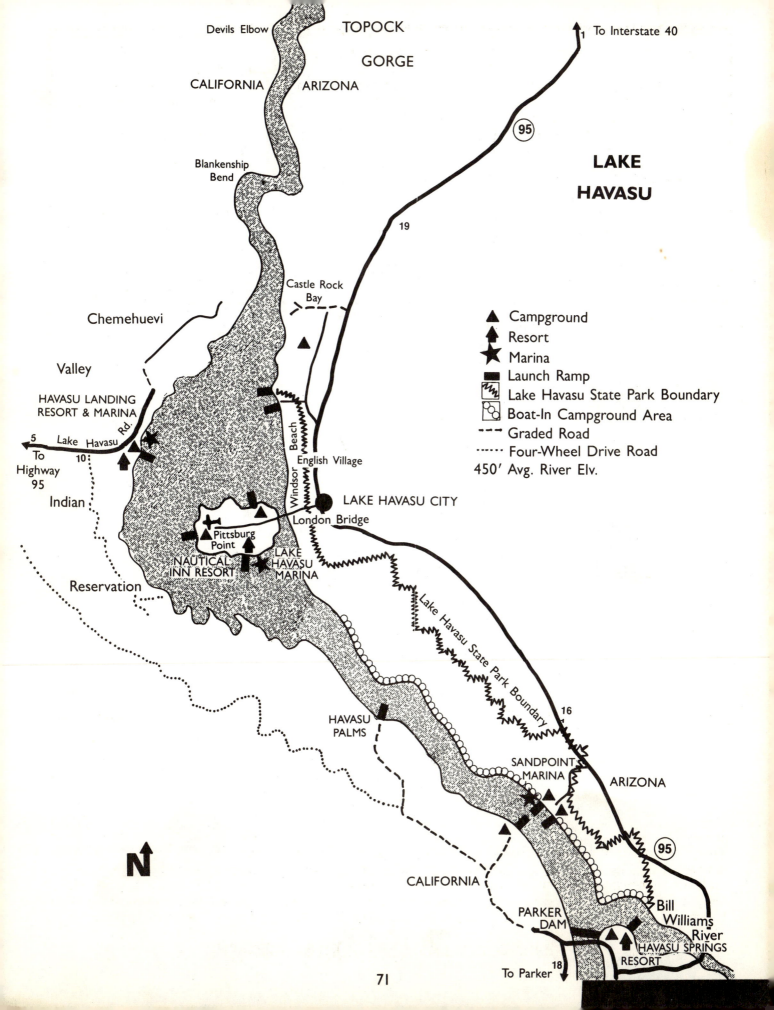

Devils Elbow

TOPOCK

GORGE

CALIFORNIA   ARIZONA

1 To Interstate 40

Blankenship Bend

95

LAKE

HAVASU

19

Chemehuevi

Castle Rock Bay

▲ Campground

▲ Resort

★ Marina

■ Launch Ramp

〰 Lake Havasu State Park Boundary

⬚ Boat-In Campground Area

--- Graded Road

⋯ Four-Wheel Drive Road

450' Avg. River Elv.

Valley

HAVASU LANDING RESORT & MARINA

5 Lake Havasu Rd.

To Highway 95

10

Indian

Windsor Beach

English Village

LAKE HAVASU CITY

Pittsburg Point

London Bridge

NAUTICAL INN RESORT

LAKE HAVASU MARINA

Reservation

Lake Havasu State Park Boundary

16

HAVASU PALMS

SANDPOINT MARINA

ARIZONA

95

N

CALIFORNIA

PARKER DAM

Bill Williams River

HAVASU SPRINGS RESORT

To Parker 18

LAKE HAVASU

Flowing out of Topock Gorge, the Colorado River becomes Lake Havasu. This 19,300 acre Lake of secluded coves, quiet inlets and open water backs up 45 miles behind Parker Dam. Its width varies from one quarter mile to over three miles, and the average depth is 40 feet. Parker Dam, completed in 1938, was financed by the Metropolitan Water District of Southern California and provides water and hydroelectric power to that nearby metropolis.

While the surrounding lands are barren mountains, the shoreline is often marshy with many coves lined with shrubs and trees. The warm, sometimes hot, clean air and low humidity lures thousands to this "Oasis" in the desert. Named after the Indian word meaning "land of the blue-green waters," Lake Havasu is a prime attraction for those wanting to enjoy its natural abundance.

Lake Havasu is one of the most popular attractions on the Colorado River. Famed for its large striped bass and warm water fishery, the Lake seldom disappoints the angler. This is an outstanding boating Lake. There are several major regattas held each year, and the sailing is excellent. Powerboating and waterskiing is outstanding with major competitions held yearly. The boat camper and houseboater will find the over 13,000 acres of Lake Havasu State Park a pleasant retreat. Swimming or just plain relaxing in the warm desert air is enjoyed by all. There are excellent support facilities and accommodations.

INFORMATION: Lake Havasu Area Chamber of Commerce, Visitor & Convention Bureau Division, 65 N. Lake Havasu Ave., Suite 2-B, Lake Havasu City, AZ 86403, Ph: 602-855-4114 or 602-453-3444

CAMPGROUNDS & RESORTS	RECREATION	OTHER
Numerous Facilities—Around the Lake See Following Pages	Power, Row, Canoe, Sail, Water-ski, Jet Ski, Windsurf & Inflatable Full Service Marinas Launch Ramps Rentals: Fishing, Power, Pontoons & Houseboats High Winds Can Be a Hazard in Fall & Spring Each Year Fishing: Catfish, Bluegill, Crappie, Largemouth & Striped Bass Swimming, Picnicking, Hiking, Backpacking, Nature Study Hunting: Waterfowl, Quail & Dove	Full Resort Facilities Airport Golf Courses Tennis Courts Boat Excursions Casino Trips Home of the London Bridge

UPPER LAKE HAVASU

This section of Lake Havasu backs up into Topock Gorge. Boaters, houseboaters and canoers share the scenic canyon of towering walls, sand bars and reed beds. Boaters should be aware that water levels fluctuate and low water can be a hazard. Castle Rock Bay is a favorite take-out point for those float boaters enjoying the River above. Havasu Landing Resort, owned by the Chemehuevi Indian Tribe, offers excellent accommodations and protected marine facilities. Located on the California side of the Lake, Havasu Landing has a passenger ferry service to English Village on the Arizona shore which leaves and returns five times daily.

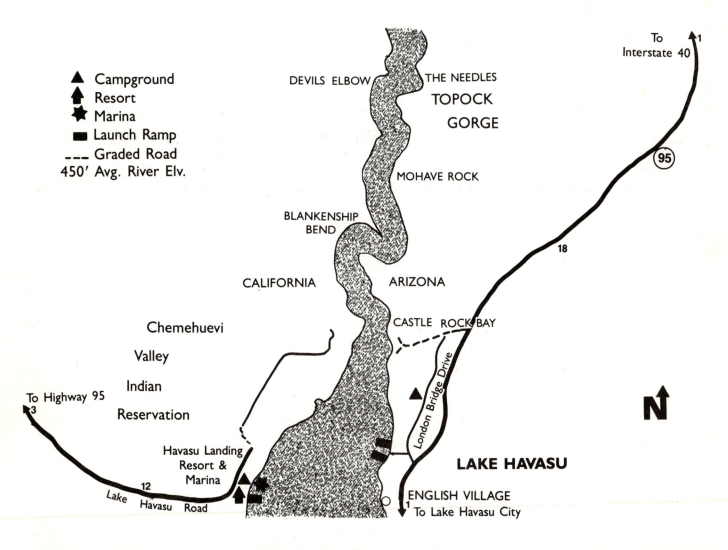

▲ Campground
🔺 Resort
★ Marina
▬ Launch Ramp
--- Graded Road
450' Avg. River Elv.

DEVILS ELBOW
THE NEEDLES
TOPOCK GORGE
MOHAVE ROCK
BLANKENSHIP BEND
CALIFORNIA
ARIZONA
CASTLE ROCK BAY
Chemehuevi Valley Indian Reservation
To Highway 95
3
To Interstate 40
95
18
London Bridge Drive
LAKE HAVASU
N
Havasu Landing Resort & Marina
12
Lake Havasu Road
ENGLISH VILLAGE
1
To Lake Havasu City

INFORMATION: Havasu Landing Resort, P.O. Box 1975, Chemehuevi Valley, CA 92363, Ph: 619-858-4593

CAMPGROUNDS & RESORTS	RECREATION	OTHER
Havasu Landing Resort—300 R.V. Sites, Full Hookups, Disposal Station, 500 Tent Sites, Cabins, Picnic Areas, Hot Showers, Snack Bar, Restaurant & Lounge, Grocery Store, Gas Station, Supplies **London Bridge KOA**—84 Tent/R.V. Sites, Full & Partial Hookups, Disposal Station, Hot Showers, Laundry, Groceries, Propane	**Havasu Landing Marina**—Full Service Protected Marina, 200 Slips, Launch Ramps, Guest Slips, Gas, Rentals: Houseboats, Fishing & Power Boats See Previous Page	Full Facilities in Needles and Havasu City

LAKE HAVASU CITY PITTSBURG POINT

Lake Havasu City is the center of the recreational opportunities on Lake Havasu. This planned resort community provides an abundance of fine accommodations, recreation and marine facilities and good restaurants. Lake Havasu's most famous attraction is historic London Bridge which was brought here, piece by piece, from London. This famous Bridge now spans the Lake between the City and the Island of Pittsburg Point. The recreational Island is a tropical paradise with sandy beaches and tall palm trees, an outstanding resort, a large campground, trailer park and full service marina. There are several major regattas and championship speed boat races held here annually.

To Havasu Landing

LAKE HAVASU

To Interstate 40

ARIZONA

95

CALIFORNIA

WINDSOR BEACH

ENGLISH VILLAGE

Beach Comber Blvd.

Pittsburg Point

Lake Havasu City

LONDON BRIDGE

THOMPSON BAY

To Parker Dam

▲ Campground
⬆ Resort
■ Picnic Site
★ Marina
▬ Launch Ramp
..... Four-Wheel Drive Road
450' Avg. River Elv.

INFORMATION: Lake Havasu Area Chamber of Commerce, 65 N. Lake Havasu Ave., Lake Havasu City, AZ 86403, Ph: 602-855-4115

CAMPGROUNDS & RESORTS

Crazy Horse Campground— 1534 Beachcomber Blvd., Lake Havasu City, AZ 86403, Ph: 602-855-2127. 741 Tent/R.V. Sites, 264 Full Hookups, 309 Partial, Disposal Stations, Ramadas, Store, Snack Bar, Swim, Beach & Waterslide, Boat Dock, 2 Launch Ramps

Lake Havasu Travel Trailer Park—P.O. Box 100, Lake Havasu City, AZ 98403, Ph: 602-855-2322. 167 R.V. Sites, Full Hookups, Swim & Therapy Pools, Boat Docks & Launch Ramp

RECREATION

Lake Havasu Marina—1100 McCulloch Blvd., Lake Havasu City, AZ 86403. Full Service Marina, 6-Lane Launch Ramp, Docks, Slips, Pumpout, Marine & Grocery Store, Bait & Tackle, Boat & Motor Repairs, Dry Storage, Waterski Boat Rentals

OTHER

Nautical Inn Resort—P.O. Box 1885, Lake Havasu City, AZ 86403, Ph: 602-855-2141. Rooms & Suites Overlooking the Lake, Private Beach & Dock, Swimming Pool, Lighted Tennis Courts, Golf Course, Restaurant & Lounge. Rental Boats: Catamarans, Windsurfers, Sailboats & Canoes, Sailing Lessons

Blue River Safaris—8 Hour Luxury Cruises. Ph: 602-763-5408.

LOWER LAKE HAVASU

Lake Havasu State Park reaches from Lake Havasu City to Parker Dam. This 13,000 acre Park along the Arizona shore provides over 160 boat-in campsites, Cattail Cove State Campground and Sand Point Marina and R.V. Park. While camping, houseboating and boating are popular attractions, Lake Havasu has long been famed for its good fishing. Perhaps the most prized gamefish is the striped bass. Imported by the Arizona Department of Fish and Game, these huge lunkers now range the entire Lake awaiting the angler. Havasu is also blessed with an ample supply of largemouth bass, crappie and channel catfish.

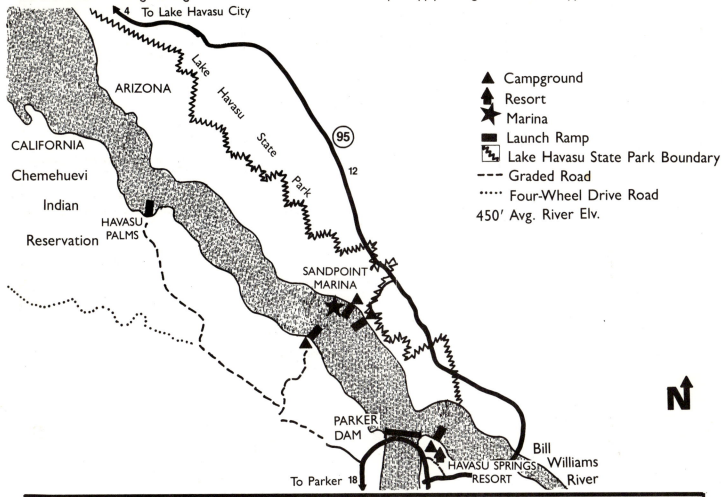

INFORMATION:
Lake Havasu State Park, 1350 McCulloch Blvd., Lake Havasu City, AZ 86403, Administration Office Ph: 602-855-7851. Cattail Cove Ranger Station, Ph: 602-855-1223

CAMPGROUNDS & RESORTS

Cattail Cove, Lake Havasu State Park—40 Tent/R.V. Sites, Water & Electric Hookups, Disposal Station, Boat Ramp

Sandpoint Marina & R.V. Park—P.O. Box 1469, Lake Havasu City, AZ 86403, Ph: 602-855-0549. 170 Tent/R.V. Sites, Full Hookups, Disposal Stations, Laundry, Snack Bar, Store, Gas Station, Playground, Swim Beach, T.V. Hookups, Full Service Marina, Gas Dock, Slips, Launch Ramp. Rentals: Fishing, Ski, Pontoon & Houseboats

RECREATION

Lake Havasu State Park—160 Boat Access Campsites, Along Arizona Shore as Shown on Map. Most have Picnic Tables, Firepots and Vault Toilets.

Black Meadow Landing—P.O. Box 98, Parker Dam 92267, Ph: 619-663-3812. 150 Tent Sites, 31 R.V. Sites, Full Hookups, Disposal Station, T.V. Hookups, Motel, Cafe, Store, Gas Station, Boat Ramp, Slips, Gas Dock. Rentals: Fishing, Pontoon Boats, Storage.

OTHER

Havasu Springs Resort—Rt. 2, Box 624, Parker, AZ 85344, Ph: 602-667-3361. 100 R.V. Sites, Full Hookups, Cable T.V., Laundry, Motel, Restaurant & Lounge, Pool, Beach, Ramp, Slips. Rentals: Fishing, Ski, Jet Ski, Camp-a-Float, Houseboats, Gas Dock, Storage, Tennis, Golf.

PARKER DAM TO PARKER

This stretch of the Colorado River between Parker Dam and Parker is commonly referred to as the "Parker Strip." An extremely popular section of the River, this area has extensive facilities, public beaches and marinas. Waterskiing is the most popular, but boating, swimming and fishing are not far behind. Another favorite recreation is inner tube floating. There are power boating and inner tube races held yearly. The fishing is good for large channel catfish, largemouth bass, crappie and bluegill, especially in Lake Moovalya. The warm, dry desert air merging with the clear blue waters lures thousands of visitors to this water wonderland.

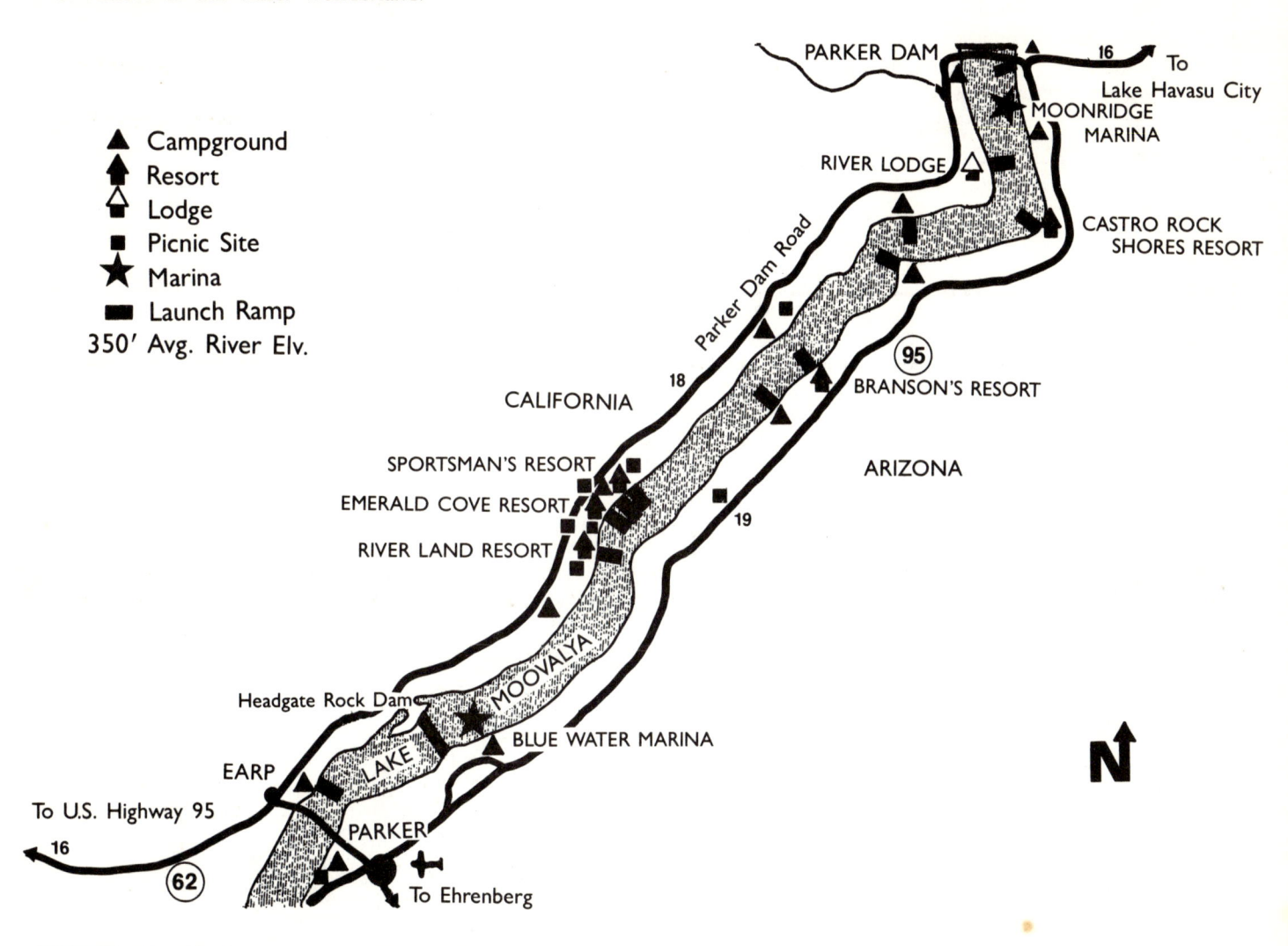

INFORMATION: Parker Chamber of Commerce, P.O. Box 627, Parker, AZ 85344, Ph: 602-669-2174.		
CAMPGROUNDS & RESORTS	**RECREATION**	**OTHER**
See Following Pages.	Open to All Boating Including Waterskiing, Racing, Tubing & Float Boating Fishing: Largemouth Bass, Channel Catfish, Crappie & Bluegill, Frogging Hunting: Cottontail Rabbit, Quail, Dove & Waterfowl Swimming Picnicking Nature Study Rockhounding	**Colorado River Indian Tribes—** Rt. 1, Box 23B, Parker, AZ 85344, Ph: 602-669-9211 **Bureau of Land Management—** Havasu Resource Area, Box 685, Lake Havasu City, AZ 86403, Ph: 602-855-8017.

PARKER DAM TO PARKER

(Continued)

The Following is a Partial List of Resorts in this Area. Contact the Parker Chamber of Commerce for Further Information:

ARIZONA

Public Campgrounds

La Paz County Park: Open Campsites to 4,500 People, No Hookups, Disposal Station, Flush Toilets, Hot Showers, Tennis Courts, Baseball Diamond, 1 Mile of Beach, 2 Boat Ramps.

> La Paz County Park
> Route 2, Box 706
> Parker, AZ 85344
> Ph: 602-667-2069

Blue Water Marina & Trailer Park (Colorado River Indian Tribes): 39 R.V. Sites, Full Hookups; 39 R.V. Sites—Dry; Restrooms, Showers, 1100 foot beach and 78 Unit Mobile Home Park.

> Blue Water Marina & Trailer Park
> Route 2, Box 430
> Parker, AZ 85344
> Ph: 602-669-5288

Buckskin Mountain State Park: 14 Tent Sites, 48 R.V. Sites with Water and Electric Hookups, 21 Cabanas on the Water, Disposal Station, Flush Toilets, Hot Showers, Groceries, Snack Bar, Gas Station, Boat Fuel, Ramp, Pontoon Boat Rentals.

> Buckskin Mountain State Park
> P.O. Box BA
> Parker, AZ 85344
> Ph: 602-667-3231

Take—Off Point Campground: 75 R.V. Sites for Self-Contained Vehicles, Pit Toilets, Boat Ramp, Fishing Area.

> Take-Off Point Campground
> Bureau of Land Management
> 3189 Sweetwater Avenue
> Lake Havasu, AZ 86403
> Ph: 602-855-8017

Private Campgrounds

Al-Do R.V. Park: 32 R.V. Sites, Full Hookups, Flush Toilets, Hot Showers, Picnic Tables, Patios.

> Al-Do R.V. Park
> P.O. Box AM
> Parker, AZ 85344

B & B Trailer Park: 20 R.V. Sites, Full Hookups, Flush Toilets, Hot Showers, Laundry, Propane.

> B & B Trailer Park
> Route 2, Box 625
> Parker, AZ 85344
> Ph: 602-667-9956

PARKER DAM TO PARKER

(Continued)

ARIZONA

Private Campgrounds (Continued)

Branson's Resort: 150 R.V. Sites, Full Hookups, Flush Toilets, Motel, Snack Bar, Boat Fuel, Ramp.

> Branson's Resort
> Route 2, Box 710
> Parker, AZ 85344
> Ph: 602-667-3346

Castle Rock Shores Resort: 110 R.V. Sites, Full Hookups, R.V. Storage, Flush Toilets, Hot Showers, Laundry, Grocery Store & Supplies, Gas, Dock, Ramp.

> Castle Rock Shores Resort
> Route 2, Box 655
> Parker, AZ 85344
> Ph: 602-667-2344

CALIFORNIA

Bermuda Palms: 89 R.V. Sites, Full Hookups, Flush Toilets, Hot Showers, Laundry, Patios, R.V. Storage, Reservations Advised.

> Bermuda Palms
> P.O. Box 353
> Earp, CA 92242
> Ph: 619-665-2784

Californian: 35 R.V. Sites, 6 Tent Sites, Full Hookups, Flush Toilets, Hot Showers, Disposal Station, Laundry, Motel, Propane, Boat Fuel, Ramp.

> Californian
> P.O. Box 127
> Parker Dam, CA 92267
> Ph: 619-663-3818

Emerald Cove Resort: 280 R.V. Sites, R.V. Storage, Flush Toilets, Hot Showers, Disposal Station, Picnic Tables, Patios, Swimming Pool, Whirlpool, Boat Ramp & Dock.

> Emerald Cove Resort
> Parker Dam, CA 92267
> Ph: 619-663-3737

Emerald Cove Desert Riviera: 20 R.V. Sites, Full Hookups, R.V. Storage, Flush Toilets, Hot Showers, Laundry, Boat Ramp and Dock.

> Emerald Cove Desert Riviera
> Star Route, Box 111
> Earp, CA 92242
> Ph: 619-663-3737

PARKER DAM TO PARKER

(Continued)

CALIFORNIA

Empire Landing Recreation Site—Bureau of Land Management: 25 Tent Sites & 50 R.V. Sites, No Hookups, Disposal Station, Flush Toilets, General Store, Boat Ramp and Docks Next Door. (Located on California Side).

> Empire Landing Recreation Site
> Bureau of Land Management
> 3189 Sweetwater Avenue
> Lake Havasu City, AZ 86403
> Ph: 602-855-8017

River Land Resort: 106 R.V. Sites, Full Hookups, R.V. Storage, Disposal Station, Flush Toilets, Hot Showers, Laundry, Groceries, Picnic Tables, Boat Fuel, Ramp.

> River Land Resort
> Star Route, Box 105
> Earp, CA 92242
> Ph: 619-663-3733

River Lodge: 300 Tent Sites, 16 R.V. Sites, Partial Hookups, Disposal Station, Flush Toilets, Hot Showers, Grocery Store, Propane, Gas, Restaurant, Boat Fuel, Ramp.

> River Lodge
> P.O. Box 57
> Parker Dam, CA 92267
> Ph: 619-663-3891

Riverview Trailer Park: 65 R.V. Sites, Full Hookups, R.V. Storage, Flush Toilets, Hot Showers, Laundry, Patios, Boat Dock, Ramp.

> Riverview Trailer Park
> P.O. Box 47
> Parker Dam Highway
> Earp, CA 92242
> Ph: 619-665-9953

Sportsman's Resort: 114 Tent Sites, 39 R.V. Sites, Full Hookups, Flush Toilets, Hot Showers, R.V. Storage, Laundry, Grocery Store, Gas, Picnic Tables, Patios, Boat Fuel, Dock, Ramp.

> Sportsman's Resort
> P.O. Box 8
> Parker Dam, CA 92267
> Ph: 619-663-3636

PARKER TO BLYTHE

The quiet gentle flow of the Colorado River entices the water sports enthusiast to this section of the River. The majority of the surrounding shoreline lies within the Colorado River Indian Reservation. Numerous Indian artifacts and handicrafts are on display at the Indian Administrative Center in Parker. The four Colorado River Tribes—Mohave, Chemehuevi, Navaho and Hopi, encourage visitors and actively support outdoor recreation in this area. Boating, waterskiing, inner tubing and canoeing are excellent. While the angler will catch an occasional striper and rainbow in the cooler months, the warm water fishing is excellent. Hunting is good, and birding can be a rewarding experience. There are numerous campgrounds and support facilities.

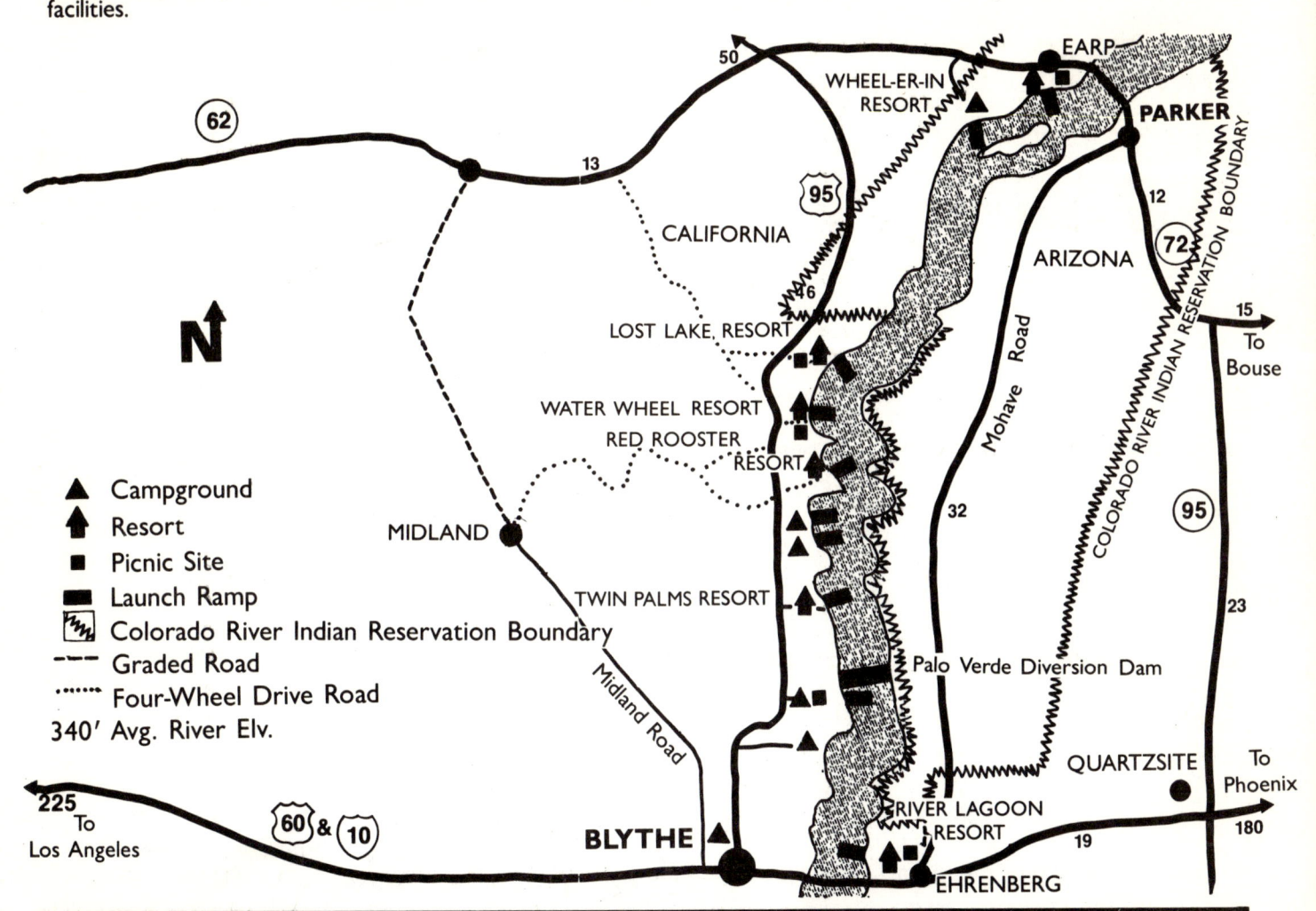

▲ Campground
⬆ Resort
■ Picnic Site
▬ Launch Ramp
〰 Colorado River Indian Reservation Boundary
--- Graded Road
····· Four-Wheel Drive Road
340' Avg. River Elv.

INFORMATION: Blythe Chamber of Commerce, 201 S. Broadway, Blythe, CA 92225, Ph: 619-922-8166

CAMPGROUNDS & RESORTS	RECREATION	OTHER
See Following Pages	Open to All Boating, Canoeing and Tubing Power Racing Fishing: Largemouth & Smallmouth & Striped Bass, Channel & Flathead Catfish, Green & Redear Sunfish, Bluegill, Crappie & Rainbow Trout Swimming & Picnicking Nature Study & Rockhounding Hunting: Cottontail Rabbits, Quail, Dove & Waterfowl Horseback Riding & Golf ORV Trails & Sand Drags	Full Facilities in Parker & Blythe **Colorado River Indian Tribes—** Tribal Administrative Center, Route 1, Box 23-B, Parker, AZ 85344, Ph: 602-669-9211

CAMPGROUNDS

Wheel-er-In Family Resort: 48 R.V. Sites, Full Hookups, Disposal Station, R.V. Storage, Flush Toilets, Hot Showers, Laundry, Grocery Store and Supplies, Picnic Tables, Patios, Boat Fuel, Ramp, Dock.

>Wheel-er-In Family Resort
>P.O. Box 46
>Earp, CA 92242
>Ph: 619-665-9952

Big River R.V. Park: 150 R.V. Sites, Full Hookups, Separate Tent Area, Disposal Station, Flush Toilets, Hot Showers, Laundry, Boat Ramp.

>Big River R.V. Park
>P.O. Box 2398
>Big River, CA 92242
>Ph: 619-665-9359

Lost Lake Resort: 300 Tent/R.V. Sites, 150 Full Hookups, Disposal Station, Separate Tent Area, Flush Toilets, Hot Showers, Laundry, Grocery Store, Gas, Picnic Tables, Patios, Boat Fuel, Ramp, Dock, 9-Hole Golf Course.

>Lost Lake Resort
>P.O. Box 6246
>Lost Lake Station
>Blythe, CA 92226
>Ph: 619-664-4413

Water Wheel Resort & Recreation Area: 100 Tent/R.V. Sites, 80 Full Hookups, R.V. Storage, Flush Toilets, Hot Showers, Laundry, Grocery Store, Picnic Tables, Boat Repairs, Boat Fuel, Ramp.

>Water Wheel Resort & Recreation Area
>HCR 20 Box 2900
>Blythe, CA 92225
>Ph: 619-922-3863

Red Rooster Resort: 12 Tent/R.V. Sites, 12 Electric & Water Hookups, Disposal Station, Hot Showers, Laundry, Deli, Grocery Store, Canoe Rentals, Boat Ramp.

>Red Rooster Resort
>HCR 20 Box 2725
>Blythe, CA 92225
>Ph: 619-922-5567

Aha-Quin Park: Trailer Park and Camp Site, General Store, Laundry, Recreation Room, Propane, Gas, Boat Ramp and Dock.

>Aha-Quin Park
>HCR 20 Box 2400
>Blythe, CA 92225
>Ph: 619-922-3604

PARKER TO BLYTHE
(Continued)

CAMPGROUNDS

Paradise Point Camp: Trailer Park and Camp Sites, General Store, Fuel, Boat Ramp.

> Paradise Point Camp
> HCR 20 Box 1750
> Blythe, CA 92225
> Ph: 619-922-8380

Twin Palms Resort: 23 Tent/R.V. Sites, Partial Hookups, Disposal Station, Flush Toilets, Hot Showers, Grocery Store, Gas, Boat Fuel & Ramp.

> Twin Palms Resort
> HCR 20 Box 1600
> Blythe, CA 92225
> Ph: 619-922-2929

Mayflower County Park: 220 Tent/R.V. Sites, Partial Hookups, Disposal Station, Flush Toilets, Hot Showers, Picnic Tables, Fire Rings, Boat Ramp, Dock.

> Mayflower County Park
> Route 1, Box 190E
> Blythe, CA 92225
> Ph: 619-922-4665

6th Avenue R.V. Park: R.V. Sites, No Tents, Gravel Ramp, Fuel Dock, Slips, Boat Rentals, Grocery Store, Bait & Tackle.

> 6th Avenue R.V. Park
> Route 1, Box 842
> Blythe, CA 92225
> Ph: 619-922-7276

River Lagoon Resort: 34 R.V. Sites, Full Hookups, R.V. Storage, Flush Toilets, Hot Showers, Laundry, Picnic Tables, Patios, Boat Ramp and Dock.

> River Lagoon Resort
> P.O. Box 3
> Ehrenberg, AZ 85334
> Ph: 602-923-7942

Burtons Trailer Park: 46 R.V. Sites, Full Hookups, Flush Toilets, Hot Showers, Laundry.

> Burtons Trailer Park
> 9395 E. Hobsonway
> Blythe, CA 92225
> Ph: 619-922-3814

BLYTHE TO OXBOW LAKE

Flowing through the Palo Verde Valley, the Colorado maintains a leisurely 6-mile per hour pace. The smooth clear water attracts a number of boaters and waterskiers, who merge with the canoeist, quietly enjoying a scenic float down the River. Boaters are cautioned to be aware of sandbars and other potential hazards, such as snags, brush, or old pilings. Oxbow Lake and Palo Verde Lagoon offer good fishing for bass, catfish and panfish. Hunting along the canals, lakes and sandbars is often excellent. There are three public parks and numerous private campgrounds on the California side of the River.

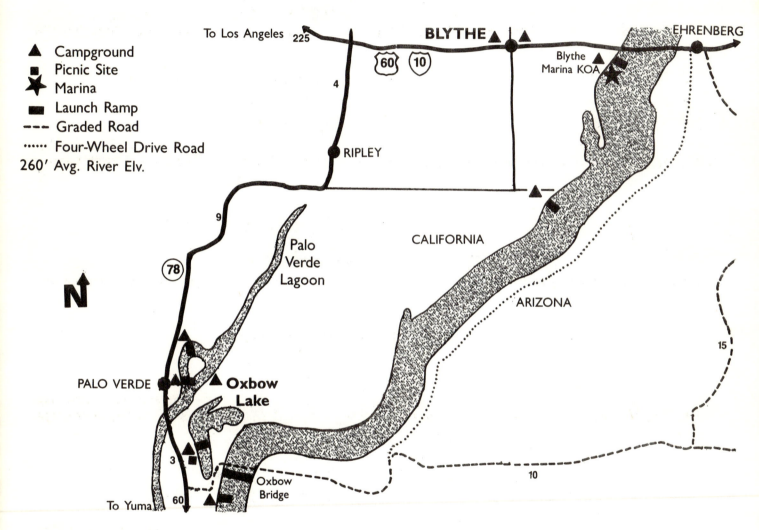

INFORMATION: Blythe Chamber of Commerce, 201 S. Broadway, Blythe, CA 92225, Ph: 619-922-8166		
CAMPGROUNDS & RESORTS	**RECREATION**	**OTHER**
See Following Pages	Open to All Boating & Canoeing Sandbars & Debris Can be a Hazard Fishing: Large & Smallmouth Bass, Channel & Flathead Catfish, Crappie, Bluegill, Green & Redear Sunfish, Perch, Frogging Hunting: Cottontail Rabbit, Dove, Quail & Waterfowl Nature Study & Rockhounding Off Road Touring & Dune Buggies	Full Facilities in Blythe **Bureau of Land Management Yuma District Office** 3150 Winsor Avenue, Yuma AZ 85364, Ph: 602-726-6300

BLYTHE TO OXBOW LAKE
(Continued)

PUBLIC CAMPGROUNDS

Palo Verde County Park: 25 Tent/R.V. Sites, No Hookups, Vault Toilets, Grocery Store, Propane, Laundry, Launch Ramp, Picnic Sites.

> Palo Verde County Park
> c/o Imperial County Parks & Recreation Dept.
> 939 S. Main Street
> El Centro, CA 92243
> Ph: 619-353-4266

Palo Verde Oxbow: 40 Tent/R.V. Sites, No Hookups, Vault Toilets, Launch Ramp.

> Palo Verde Oxbow
> c/o BLM
> 3150 Winsor Avenue
> Yuma, AZ 85364
> Ph: 602-726-6300

PRIVATE CAMPGROUNDS

Arp's Trailer Park: 14 R.V. Sites, Full Hookups, Flush Toilets, Hot Showers, Laundry, Propane, Boat Rentals, Ramp.

> Arp's Trailer Park
> Box 328
> Palo Verde, CA 92266
> Ph: 619-854-3195

Burton's Trailer Park: 46 R.V. Sites, Full Hookups, Flush Toilets, Hot Showers, Laundry.

> Burton's Trailer Park
> 9395 E. Hobsonway
> Blythe, CA 92225
> Ph: 619-922-3814

McIntyre County Park: 120 Tent Sites, 140 R.V. Sites, Electric & Water Hookups, Disposal Station, Flush Toilets, Hot Showers, General Store, Propane, Boat Fuel, Ramp, Canoe Rentals.

> McIntyre County Park
> Route 2, Box 179D
> Blythe, CA 92225
> Ph: 619-922-8205

PRIVATE CAMPGROUNDS

(Continued)

Riviera Blythe Marina: 200 Tent/R.V. Sites, Full Hookups, Disposal Station, Flush Toilets, Hot Showers, General Store, Laundry, Propane, Gas, Heated Swimming Pool, Jacuzzi, Game Room, Boat Fuel, Dock Ramp, Reservations Advised.

> Riviera Blythe Marina
> 14100 Riviera Drive
> Blythe, CA 92225
> Ph: 619-922-5350

Mitchell Camp: 80 Tent/R.V. Sites, Full Hookups, Disposal Station, Flush Toilets, Hot Showers, General Store, Laundry, Boat Ramp.

> Mitchell Camp
> P.O. Box 111
> Palo Verde, CA 92266
> Ph: 619-854-3266

Sandbar Trailer Park: 4 R.V. Sites, Electric Hookups, Flush Toilets, Hot Showers, Restaurant, Propane.

> Sandbar Trailer Park
> P.O. Box 720
> Palo Verde, CA 92266
> Ph: 619-854-3414

Valley Palms Park: 37 R.V. Sites, Full Hookups, Disposal Station, Flush Toilets, Showers, Laundry.

> Valley Palms Park
> 8401 E. Hobsonway
> Blythe, CA 92225
> Ph: 619-922-7335

OXBOW TO MARTINEZ LAKE

Many canoe and float parties begin their scenic trip downriver at Oxbow Bridge. The flat moving River flows through the Cibola and Imperial National Wildlife Refuges which provide a scenic environment of narrow coves, inlets and hidden Lakes. Fishing is excellent. This is a hunter's paradise of abundant game and waterfowl. Birding and photography are popular. Waterskiing is only permitted in designated sections of the River (see Map). Overnight camping within the Refuge is not permitted. The Picacho State Recreation Area has campsites for the boater and traveler. This remote, quiet section of the River is ideal for those who enjoy a natural environment.

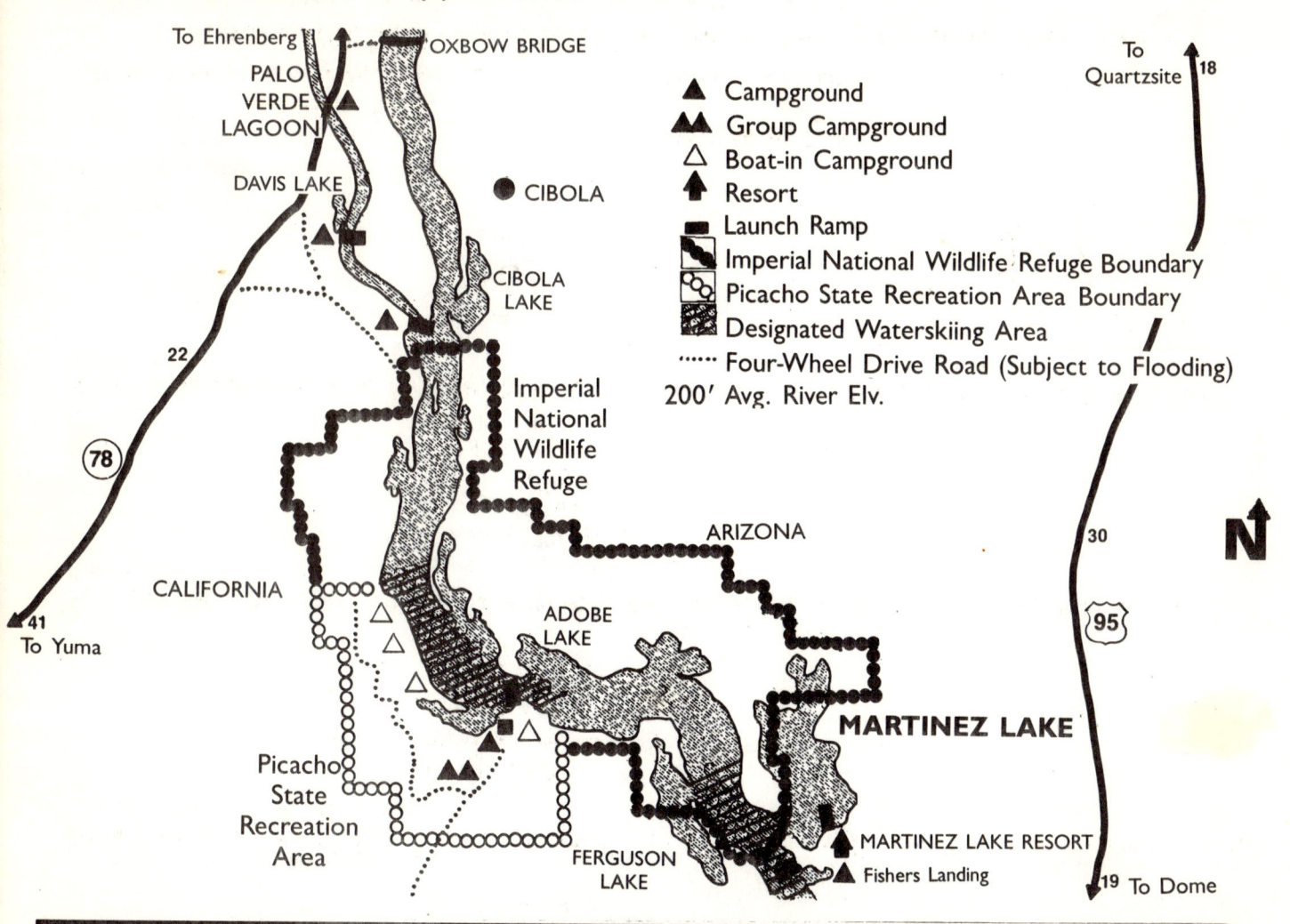

INFORMATION: Refuge Manager, P.O. Box 72217, Martinez Lake, AZ 85365, Ph: 602-783-3371

CAMPGROUNDS & RESORTS	RECREATION	OTHER
See Following Page	Open to All Boating, But Limited by Sand Bars and In-Water Hazards Waterskiing is Prohibited in Wildlife Refuge Except as Noted on Map Fishing: Large and Smallmouth Bass, Channel & Flathead Catfish, Crappie, Bluegill & Sunfish Hunting: Cottontail Rabbit, Quail, Dove & Waterfowl Birding & Nature Study Picnicking & Hiking Swimming	Full Facilities in Blythe and Yuma ORVs & Dune Buggies Nearby. Not Permitted in Picacho State Recreation Area. **Imperial Sand Dunes**—c/o Bureau of Land Management, El Centro Resource Center, El Centro, CA 92243, Ph: 619-352-5842 **OXBOW TO MARTINEZ LAKE** Bureau of Land Management— Yuma District Office, 3150 Winsor Avenue, Yuma, AZ 85364, Ph: 602-726-6300.

(Continued)

Coco Palms Mobile Home Park: 25 Tent/R.V. Sites, Full Hookups, Disposal Station, Flush Toilets, Hot Showers, Laundry.

> Coco Palms Mobile Home Park
> P.O. Box 278
> Palo Verde, CA 92266
> Ph: 619-854-3391

Mitchell Camp: 80 Tent/R.V. Sites, Full Hookups, Disposal Station, Flush Toilets, Hot Showers, Grocery Store, Laundry, Boat Ramp.

> Mitchell Camp
> P.O. Box 111
> Palo Verde, CA 92266
> Ph: 619-854-3266

Fishers Landing: 162 R.V. Sites, Full Hookups, Flush Toilets, Hot Showers, Laundry, Grocery Store, Restaurant, Gas, Propane, Boat Fuel, Rentals, Ramp.

> Fishers Landing
> Star Route 4
> Box 45
> Yuma, AZ 85365
> Ph: 602-783-6513

Martinez Lake Resort: 25 R.V. Sites, Full Hookups, Flush Toilets, Hot Showers, Laundry, Motel, Restaurant, Gas, Boat Fuel, Rentals, Ramp.

> Martinez Lake Resort
> P.O. Box 72245
> Martinez Lake, AZ 85365
> Ph: 602-783-9589

Walter's Camp: 50 Tent, 60 R.V. Sites, Flush Toilets, Hot Showers, Store, Snack Bar, Gas, Propane, Boat Fuel and Ramp.

> Walter's Camp
> P.O. Box 31
> Palo Verde, CA 92266
> Ph: 619-854-3322

PUBLIC CAMPGROUNDS

Picacho State Recreation Area: P.O. Box 1207, Winterhaven, CA 92283

> **Taylor Lake Boat-In Campground:** 4 Tent Sites, Pit Toilets, Shade Ramadas, Picnic Tables.
> **Paddlewheeler Boat-In Campground:** 3 Tent Sites, Vault Toilets
> **Picacho State Recreation Area Campground:** 5 Tent/R..V. Sites, No Hookups, Group Campground to 60 People, Group Boat-In Camp Area, Picnic Sites, Boat Ramp.

All of the above Campgrounds are reached via an 18 Mile Dirt Road
Subject to flooding during high water.

HOT WEATHER EXPLORING

The Colorado River flows through some of the most beautiful and scenic countryside in America. Exploring this sometimes arid environment can be a rewarding experience. Yet, caution is advised. Mother Nature is a demanding parent, but if you follow her rules, the benefits are many. Desert heat can create serious medical problems, the least of which is sunburn.

Heat cramps, heat exhaustion and heat stroke are a constant threat. It is the wise hiker who has educated himself in the symptoms and emergency treatment of these heat-related injuries. Never venture into this arid country without you or a member of your group knowing basic first aid procedures.

There are some simple rules to follow when in the desert. Always have and drink plenty of water — 1 gallon per day. Protect your body by wearing light, loose fitting clothing, and cover your head with a ventilated hat or baseball-type cap. A sun screen preparation on exposed skin is helpful, and protect your eyes with sun glasses and a visor. Bring and eat high energy foods rich in carbohydrates. If you are taking medication or are under doctor's care, consult him before venturing into the desert.

While being in the best of shape is an asset, it does not preclude heat injury. You must be acclimated. A rise in humidity can increase your body temperature. Always be in touch with your body and don't overexert it. Never hike or travel during the hot midday hours. Always travel during the cooler morning and late evenings, and rest in the shade during the hot periods. If an emergency occurs, find some shade, relax, drink and get help when necessary.

YUMA AREA

Flowing through open desert periodically accented by barren mountains, the Colorado meanders into backwater lagoons and marshy sloughs. Much of its water is diverted into the All American Canal at Imperial Dam. Many "snowbirds," both fowl and human, flock to this warm winter area. The campgrounds are busy and the birds are plentiful. The hunting is good. The warm water fishery is excellent, especially for huge flathead catfish. Boating is somewhat limited to craft with a shallow draft except as noted in the graph below. The nearby Imperial Sand Dunes Recreation Area offers the largest mass of sand dunes in California and provides a haven for off-road vehicle enthusiasts.

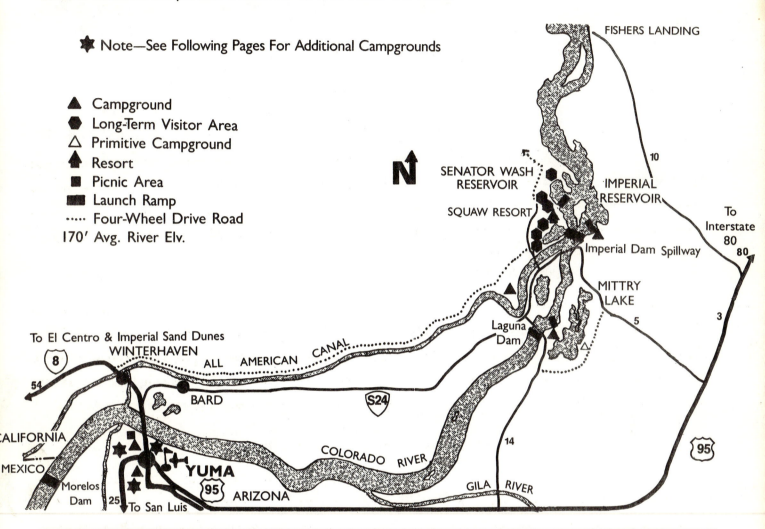

★ Note—See Following Pages For Additional Campgrounds

▲ Campground
⬣ Long-Term Visitor Area
△ Primitive Campground
⬆ Resort
■ Picnic Area
▬ Launch Ramp
····· Four-Wheel Drive Road
170' Avg. River Elv.

INFORMATION: Yuma County Chamber of Commerce, Box 230, Yuma, AZ 85364, Ph: 602-782-2567

CAMPGROUNDS & RESORTS

See Following Pages

Most Campgrounds Offer Seasonal Leases During Winter Months. Advance Reservations Advised.

General Camping within the Jurisdiction of the Bureau of Land Management is limited to 14 days except in the Designated Long Term Visitors Areas as shown on Map. Contact the BLM for Permits in LTVAs.

RECREATION

Open to All Boating, but Water-skiing is Limited to Behind Imperial Dam and Senator Wash Reservoir

Fishing: Large & Smallmouth Bass, Channel & Flathead Catfish, Crappie, Pan Fish

Hunting: Cottontail Rabbits, Quail, Dove & Waterfowl

Swimming & Picnicking

Hiking & Nature Study

ORV's & Dune Buggies

Rockhounding

OTHER

Bureau of Land Management Yuma District Office—3150 Winsor Avenue, Yuma, AZ 85364, Ph: 602-726-6300

California District Office—1695 Spruce Street, Riverside, CA 92507, Ph: 714-351-6394

El Centro Resource Area—333 S. Waterman Avenue, El Centro, CA 92243, Ph: 619-352-5842

YUMA AREA

(Continued)

PUBLIC CAMPGROUNDS

Squaw Lake Recreation Site, Bureau of Land Management: 3150 Winsor Ave., Yuma, AZ 85364, Ph: 602-726-6300

160 Tent/R.V. Sites, No Hookups, Disposal Station, Cold Showers, Flush Toilets, Nature Trails, Swim Beach, Boat Ramp.

Laguna Dam South Recreation Site: Route 1, Laguna Dam South, Winterhaven, CA 92283, Ph: 619-572-0798

475 Tent/R.V. Sites, No Hookups, Disposal Station, Cold Showers, Flush Toilets.

PRIVATE CAMPGROUNDS

This is a **Partial** List. We have shown these Campgrounds as a Sample
of the Extensive Facilities in this Area of the Colorado River.
Contact the Chamber of Commerce for Further Details.

Atlasta Trailer Park: 415 S. May St., Yuma, AZ 85365, Ph: 602-783-1925.

113 R.V. Sites, Full Hookups, R.V. Storage, Hot Showers, Flush Toilets, Disposal Station, Laundry, Patios.

Blue Sky R.V. Park: 10247 Fwy. I-8, Yuma, AZ 85365, Ph: 602-343-1444.

155 R.V. Sites, Full Hookups, R.V. Storage, Hot Showers, Flush Toilets, Laundry, Patios, Whirlpool.

Caravan Oasis Adult R.V. Park: 10500 E. Frontage Rd., Yuma, AZ 85365, Ph: 602-342-1480.

620 R.V. Sites, Full Hookups, R.V. Storage, Hot Showers, Flush Toilets, Handicapped Facilities, Laundry, Patios, Swimming Pool, Whirlpool.

Country Roads R.V. Village: 5707 E. Hwy. 80, Yuma, AZ 85364, Ph: 800-222-3939 or 602-344-8910.

1,302 R.V. Sites, Full Hookups, Laundry, 3 Swimming Pools & Jacuzzis, 4 Saunas, Tennis Court, Recreation Facilities.

8th Street Mobile Home & R.V. Park: 1640 West 8th St., Yuma, AZ, Ph: 602-782-1789.

89 R.V. Sites, Full Hookups, R.V. Storage, Hot Showers, Flush Toilets, Laundry, Patios.

Foothill Village R.V. Park: 12705 E. Frontage Rd., Yuma, AZ 85365, Ph: 602-342-1030.

188 R.V. Sites, Full Hookups, Hot Showers, Flush Toilets, Disposal Station, Laundry, Patios & Picnic Tables, Swimming Pool, Whirlpool.

Friendly Acres R.V. Trailer Park: 2779 West 8th St., Yuma, AZ 85365, Ph: 602-783-8414.

287 R.V. Sites, Full Hookups, Hot Showers, Flush Toilets, Disposal Station, Laundry, Patios & Picnic Tables, Propane, Swimming Pool, Adults.

Gila Mountain R.V. Park: 12325 E. Frontage Rd., Yuma, AZ 85365, Ph: 602-342-1310

275 R.V. Sites, Full Hookups, Hot Showers, Flush Toilets, Disposal Station, Laundry, R.V. Supplies, Propane, Patios & Picnic Tables, Swimming Pool, Whirlpool, Adults, Open October 1 to June 1.

Hidden Cove Trailer Park and Marina: 2450 West Water St., Yuma, AZ 85365, Ph: 602-783-3534.

> 110 R.V. Sites, Full Hookups, Separate Tent Area, Hot Showers, Flush Toilets, Laundry, Patios, River Swimming, Boat Ramp, Dock, Adults, Open October 1 to March 30.

Imperial Oasis: Star Route #4, Yuma, AZ 85365, Ph: 602-783-4171. (Located on Shore of Imperial Reservoir)

> 25 Tent Sites, 240 Sites, Hot Showers, Flush Toilets, Laundry, General Store, Groceries, Restaurant, Propane, Gas, Boat Fuel, Ramp, Boat Rentals.

Lucky Park Del Sur: 5790 West 8th St., Yuma, AZ 85365, Ph: 602-783-7201.

> 51 R.V. Sites, Full Hookups, Disposal Station, Hot Showers, Flush Toilets, Patios, Laundry, Cable TV.

Sheltering Palms Trailer Ranch: 2545 West 8th St., Yuma, AZ 85365, Ph: 602-782-2910.

> 160 R.V. Sites, Full Hookups, R.V. Storage, Hot Showers, Flush Toilets, Laundry, Patios, Adults.

Spring Garden Trailer Park: 3550 West 8th St., Yuma, AZ 85365, Ph: 602-342-9333.

> 140 R.V. Sites, Full Hookups, Disposal Station, Hot Showers, Flush Toilets, Laundry, Patios, Whirlpool, Adults.

Sun Glo Mobile Home & R.V. Park: 510 East Highway 80, Yuma, AZ 85365, Ph: 602-726-0160.

> 137 R.V. Sites, Full Hookups, Disposal Station, Hot Showers, Flush Toilets, Laundry, Patios, Adults.

Sun Vista R.V. Resort: 7201 E. Hwy. 80, Yuma, AZ 85365, Ph: 602-783-9443.

> 1,230 R.V. Sites, Full Hookups, Disposal Station, Hot Showers, Flush Toilets, 2 Indoor & Outdoor Swimming Pools, Jacuzzis, Laundry, T.V. Room, Ball Room, Craft Room.

Yuma Mesa R.V. Park: 5990 East Highway 80, Yuma, AZ 85365, Ph: 602-344-3369.

> 184 R.V. Sites, Full Hookups, R.V. Storage, Hot Showers, Flush Toilets, Laundry, Patios, Swimming Pool, Adults.

Yuma Overnite Trailer Park: 201 West 28th St., Yuma, AZ 85365, Ph: 602-355-1012.

> 70 R.V. Sites, Full Hookups, R.V. Storage, Hot Showers, Flush Toilets, Disposal Station, Laundry, Patios.

UNITED STATES/MEXICO BORDER TO THE SEA OF CORTEZ

The mighty Colorado enters Mexico at San Luis flowing into the Delta below. Nearing the end of its journey, the River's waters are met by the tidal surge of the Gulf of Lower California only reaching the Sea of Cortez at low tides. Exhausted after its turbulent adventure, it finally yields its throne to the Sea just 70 miles below the border. Gentle sea breezes, broad sandy beaches and over 800 species of game fish invite the visitor to this area. Facilities are limited and roads are sparse, so a four-wheeler is often advised.

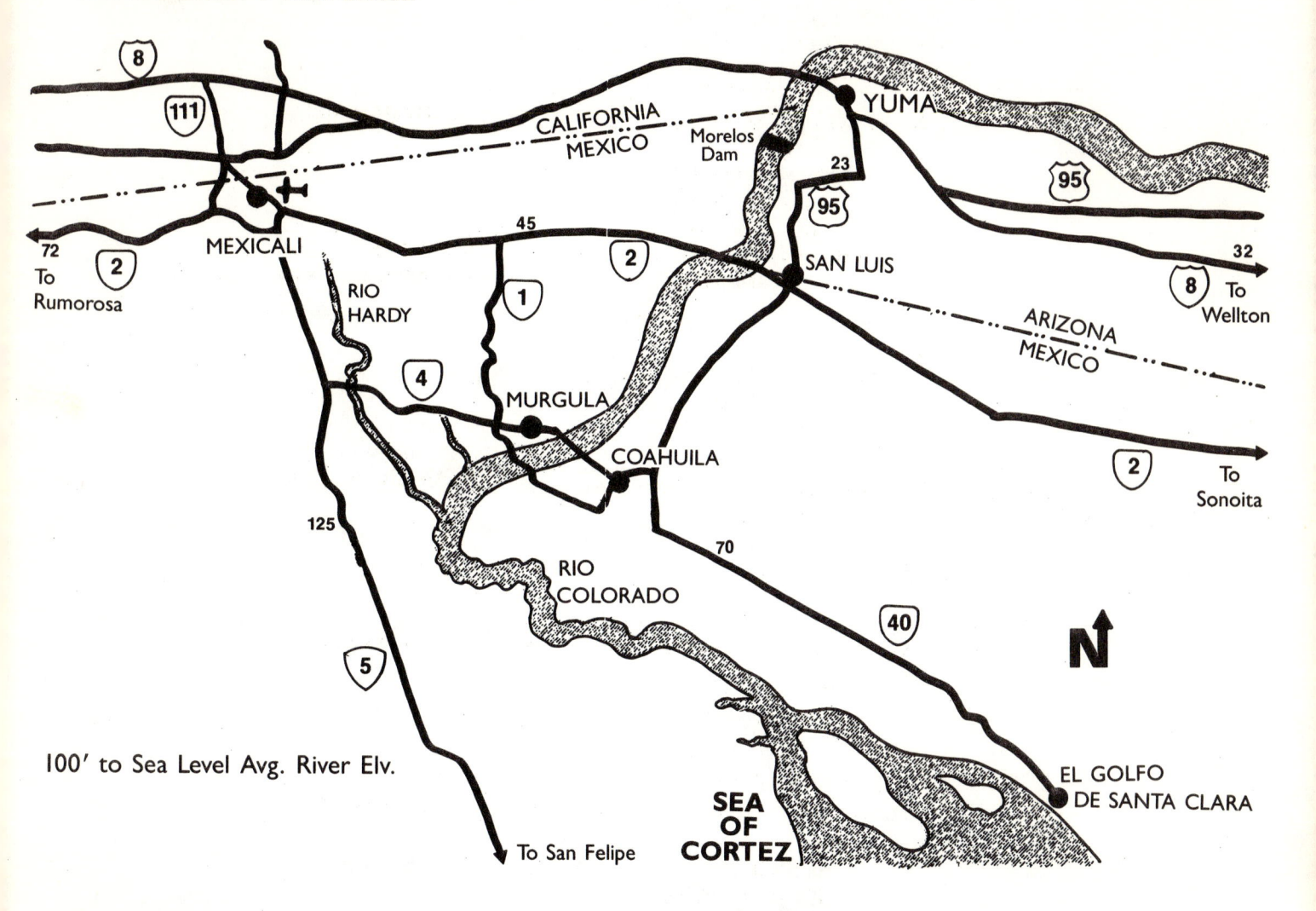

100' to Sea Level Avg. River Elv.

INFORMATION: Yuma County Chamber of Commerce, P.O. Box 230, 377 Main St., Yuma, AZ 85364, Ph: 602-782-2567		
CAMPGROUNDS & RESORTS	**RECREATION**	**OTHER**
Limited to Rugged Beach Camping. Bring Your Own Drinking Water.	Boating Limited Except in the Sea of Cortez Fishing: Over 800 Species in the Gulf of California, Clam Digging Hunting: Quail, Dove & Waterfowl Swimming: Beaches 4-Wheel Driving Nature Study & Exploring	Full Facilities in Yuma, Mexicali and San Luis, Mexico **Mexicali Tourism and Convention Bureau—** Blvd. Lopez Mateos y Calle Camelias, Mexicali, Ph: 706-567-2376 or 706-567-2561 **State Tourism Office—** Ph: 706-562-4391 or 706-562-9795

GOVERNMENT INFORMATION CONTACTS FOR SECTIONS OF THE COLORADO RIVER

COLORADO — UTAH

Glenwood Canyon:

USDA Forest Service
Eagle Ranger District
White River National Forest
P.O. Box 720
Eagle, CO 81631
Ph: 303-328-6388

Pumphouse to State Bridge:

Bureau of Land Management
Kremmling Resource Area
P.O. Box 68
Kremmling, CO 80549
Ph: 303-724-3437

State Bridge to Dotsero:

Bureau of Land Management
Glenwood Springs Resource Area
P.O. Box 1009
Glenwood Springs, CO 81602
Ph: 303-945-2341

**Loma to Westwater —
Dolores River**
(Gateway to Utah Line)

Bureau of Land Management
Grand Junction Resource Area
764 Harrison Dr.
P.O. Box 1509
Grand Junction, CO 81501
Ph: 303-243-6552

Cataract Canyon
(Including Green River)

Superintendent
Canyonlands National Park
446 South Main
Moab, UT 84532
Ph: 801-259-7164

Hittle Bottom to Sandy Beach

Bureau of Land Management
Grand Resource Area
P.O. Box M
Moab, UT 84532
Ph: 801-259-8193

GOVERNMENT INFORMATION CONTACTS FOR SECTIONS OF THE COLORADO RIVER

ARIZONA — CALIFORNIA

Grand Canyon

River Permits Office
National Park Service
Grand Canyon National Park
P.O. Box 129
Grand Canyon, AZ 86023
Ph: 602-638-7843

Blythe to Imperial Dam

Bureau of Land Management
P.O. Box 5680
Yuma, AZ 85364
Ph: 602-726-6300

Laguna Dam

Bureau of Indian Affairs
P.O. Box 1591
Yuma, AZ 85364
Ph: 714-572-0248

Quechan Tribe
P.O. Box 1352
Yuma, AZ 85364
Ph: 714-572-0820

Westwater Canyon

Bureau of Land Management
Grand Resource Area
P.O. Box M
Moab, UT 84532
Ph: 801-259-8193

Dolores River
(Gateway to Colorado River)

Bureau of Land Management
P.O. Box M
Moab, UT 84532
Ph: 801-259-8193

San Juan River
(Montezuma Creek to
Clay Hills Crossing)

Bureau of Land Management
San Juan Resource Area
P.O. Box 7
Monticello, UT 84535
Ph: 801-587-2201

STATE AGENCIES

In addition to those Agencies listed in the text of this book, the State Agencies shown below can provide valuable information.

They regulate fishing and hunting, commercial licensing, state boating safety and the use of State lands:

ARIZONA

Director, Fish and Game Department
P.O. Box 9099
Phoenix, AZ 85068
Ph: 602-628-5376

Arizona State Parks
Natural Areas and Trails Co-ordinator
1688 W. Adams
Phoenix, AZ 85068
Ph: 602-255-4174

CALIFORNIA

Department of Fish and Game
Resource Building
1416 - 9th St.
Sacramento, CA 95814
Ph: 916-445-3531

California Parks and Recreation Dept.
915 Capitol Mall
Sacramento, CA 95814
Ph: 916-445-6477

COLORADO

Chief, Law Enforcement
Colorado Division of Wildlife
6060 Broadway
Denver, CO 80216
Ph: 303-825-1192

State Boating Safety Officer
Colorado Division of Parks and
Outdoor Recreation
Chatfield Office Center
13787 S. Highway 85
Littleton, CO 80125
Ph: 303-795-6954

NEVADA

Nevada Department of Wildlife
P.O. Box 10678
Reno, NV 89520
Ph: 702-789-0500

UTAH

Utah Division of Parks and Recreation
Boating Chief
1596 W. North Temple
Salt Lake City, UT 84116
Ph: 801-533-6011

Utah Travel Council
Council Hall/Capitol Hill
Salt Lake City, UT 84114
Ph: 801-533-5682

Utah Division of Wildlife Resources
1596 West North Temple
Salt Lake City, UT 84116
Ph: 801-533-9333

Assistant Director of Auxillary
U.S. Coast Guard
Salt Lake City, UT 84111
Ph: 801-524-5155

RIVER RUNNING

River running is one of the fastest growing water sports in America. There is no River in the Northern Hemisphere offering such a variety of environments as the Colorado. While the novice boater may safely enjoy the quiet sections of the River, only the experienced white water boater should attempt the more demanding rapids.

Commercial River Runners offer the complete adventure of the Colorado and its tributaries for everyone from 7 to 80. Many of these commercial outfitters provide dedicated crews, well versed in white water boating, wildlife, geology and River history. The meals provided are often outstanding. These outfitters are an important part of the Colorado River experience.

Many sections of the River limit the traffic to preserve its natural state. The Grand Canyon, Canyonlands and Cataract Canyon limit the number of commercial outfitters. There are specific rules regulating various sections of the River. Each page of this book provides the address and phone number of the governing agencies who are able to provide you with specific information regarding licensed outfitters and regulations. In addition, the various Chamber of Commerce and Tourism Offices can provide you with a list of outfitters.

Important General Information sources are listed below:

Colorado River Outfitters Association
7600 E. Arapahoe Road, Suite 114
Englewood, CO 801112

Western River Guides Association, Inc.
4260 East Evans Ave.
Denver, CO 80222
Ph: 303-758-5715

Western River Guides Association, Inc.
1424 Lancelot Dr.
Provo, UT 84601
Ph: 801-373-1223

Utah Travel Council
Tour Guide to Utah
Council Hall/Capitol Hill
Salt Lake City, UT 84114
Ph: 801-533-5681

Grand County Travel Council
Tourism Services
805 N. Main St.
Moab, UT 84532
Ph: 801-259-8825

COMMERCIAL OUTFITTERS

We regret that we are unable to furnish a complete list of Commercial Outfitters. Many new Companies are starting, others are leaving. We do not have the space to list all of them, so the following list was selected at random. Other Outfitters can be found through the various informational centers listed on the previous page.

Adventure Bound, Inc.
122 S. 8th St.
Grand Junction, CO 81501
U.S. Toll Free Ph: 1-800-525-7084
Colorado Ph: 1-800-332-1400
       or   303-245-0024

Runs a Variety of Rivers on the Western Slope Including Cataract and Westwater Canyons.

Arizona River Runners, Inc.
P.O. Box 2021
Marble Canyon, AZ 86036
Ph: 602-355-2223

Runs the Grand Canyon.

Black Canyon Adventures
Gold Strike
Boulder City, CO 89005
Toll Free Ph: 1-800-634-6579
       or   702-384-1234

Runs the Black Canyon below Hoover Dam.

Griffins Colorado Adventures, Inc.
Central Reservations Office
P.O. Box 388
Steamboat Springs, CO 80477

Runs a Variety of Rivers Including the Upper Colorado, Westwater and Cataract Canyons.

Outlaw River Expeditions, Inc.
P.O. Box 790
Moab, UT 84532

Runs the Green and Colorado Including Cataract, Desolation and Westwater Canyons.
Ph: 801-259-8241

Rock Gardens Raft Trips
1308 County Rd. 129
Glenwood Springs, CO 81601
Ph: 303-945-6737

Runs Glenwood Canyon.

Ken Sleight
Box 1270
Moab, UT 84532
Ph: 801-259-8575

Runs the San Juan, Provides Boat Trips in Glen Canyon. Pack Trips and Trail Rides, Archeological and Historical Trips

COMMERCIAL OUTFITTERS

(Continued)

Tag-A-Long Tours
P.O. Box 1206
452 North Main Street
Moab, UT 84532
Ph: 801-259-8946

Provides a Variety of Rafting and Jeeping Safaris in Canyonlands, Grey, Desolation and Westwater Canyons.

Tex's River Expeditions
Box 67
Moab, UT 84532
Ph: 801-259-5101

Runs Canyonlands and Cataract Canyon. Canoe Trips into Canyonlands.

Del Webbs Wilderness River Adventures
P.O. Box 717
Page, AZ 86040
Ph: 602-645-3279

Runs 3- to 7-Day Trips Throughout the Grand Canyon.

Wild & Scenic Canyonland Tours
P.O. Box 460
Flagstaff, AZ 86002
Ph: 602-774-7343

Runs a Variety of Tours Including the Grand and Cataract Canyons. Also Provides Photographic Tours Through Arches, Navaholands and Glen Canyon.

RIVER FLOW INFORMATION

Colorado River Basin	801-539-1311
Colorado	303-371-7739
California	916-322-3327

RAPIDS GRAPH

GENERAL	Colorado, Cataract (NPS)	Colorado, Hittle Bottom	Colorado, Grand Canyon (NPS)	Colorado, Blyth/Imperial	Colorado, Lower, Laguna Dam	Colorado, Westwater	Dolores, Gateway to Colo. River	Gila, Lower (BLM)	Green, Cataract (NPS)	Green, Desolation, Gray	Green, Labyrinth Canyon	San Juan, Montezuma to Clay Hills	Colorado, Glenwood Canyon	Colorado, Pumphouse to State Bridge	Colorado, State Bridge to Dotsero	Colorado, Loma to Westwater
Period Runnable																
Spring	x	x	x	x	x	x	x	x	x	x	x	x	x	x	x	x
Summer	x	x	x	x	x	x	x	x	x	x	x	x	x	x	x	x
Fall	x	x	x	x	x	x			x	x	x	x	x		x	x
Winter			x	x	x							x				
Difficulty																
I-II		x	x	x	x			x	x		x	x	x	x	x	x
III-IV	x		x			x	x			x		x				
V +			x													
Hazardous High Flows	x	x					x	x					x			
Limiting Low Flows			x					x			x	x			x	x
Controlled Flow			x	x	x				x	x		x		x	x	x
Trip Length—Days	3-7	1	3-18	1-5	1-3	1-3	1-3	1-2	3-7	3-6	4	1-6	1	1-3	1-3	1-3
Shuttle Service	x	x	x			x			x			x	x			
COMMERCIAL																
Float Boat—Motor	x	x	x			x	x		x	x	x	x				x
Float Boat—Non-motor	x	x	x			x	x		x	x	x	x	x	x	x	x
Small Craft—Kayak, etc.	x	x	x	x		x	x		x	x	x	x		x	x	x
NON-COMMERCIAL																
Permits Required	x		x			x	x		x	x		x				
Reservations Required	x		x			x			x	x		x				
Restrictions on Type Craft	x		x						x	x						

RAPID RATINGS

There are two systems of Rapid Ratings, the International and Western (or American) Scales. The International Scale is generally accepted as the standard.

INTERNATIONAL SCALE	WESTERN SCALE	DESCRIPTION
	0	**Flat Water.**
I	1,2	**Easy.** Waves Small, Passages Clear, No Serious Obstacles.
II	3,4	**Medium.** Rapids of Moderate Difficulty with Passages Clear. Requires Experience plus Fair Outfit and Boat.
III	5,6	**Difficult.** Waves Numerous, High, Irregular, Rocks, Eddies, Rapids with Passages Clear though Narrow, Requires Expertise in Maneuver, Scouting Usually Needed, Requires Good Operator and Boat.
IV	7,8	**Very Difficult:** Long Rapids, Waves Powerful, Irregular, Dangerous Rocks, Boiling Eddies, Passages Difficult to Scout, Scouting Mandatory First Time, Powerful and Precise Maneuvering Required. Demands Expert Boatman and Excellent Boat and Outfit.
V	9,10	**Extremely Difficult:** Exceedingly Difficult, Long and Violent Rapids, Following Each Other Almost Without Interruption, Riverbed Extremely Obstructed, Big Drops, Violent Current, Very Steep Gradient, Close Study Essential but Often Difficult, Requires Best Man, Boat and Outfit Suited to the Situation. All Possible Precautions Must be Taken.
VI or U	U	**Unrunnable.**

PETS

While pets are not allowed at some facilities, they are normally welcome at most campgrounds, parks and recreation areas. Often a nominal fee is charged, and there are some general rules to follow. Proof of a current rabies vaccination, a current license and a leash, no longer than eight feet, are usually required. Pets are not allowed to contaminate the water, nor are they allowed into Wilderness areas, on trails or swimming beaches.

ENDANGERED FISH

In the extreme environment of the Colorado River drainage, nearly 70% of the native fish are found nowhere else in the world. Four of these unique fishes are now in danger of extinction. These fishes have developed bizarre shapes, such as humped or ridged backs, winged-like fins and streamlined bodies to help them navigate the River during high water. If you catch one of these endangered species, return it to the river. All are protected by Federal and State laws.

The Colorado Squawfish is the largest minnow in North America reaching weights of 50 to 80 pounds. Once a source of food for early cultures, today it has virtually disappeared from the Lower Basin and is rare in the Upper Basin.

The Humpback Chub is only rarely found in the Little Colorado section of the Grand Canyon. Usually less than 13 inches long, scientists fear it may become extinct before its biology is understood.

ENDANGERED FISH
(Continued)

Bonytail Chub is occasionally found in Desolation Canyon on the Lower Green River. They are probably extinct in the Lower Colorado.

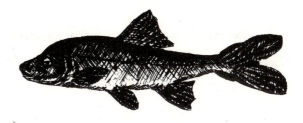

Razorback Sucker is one of the largest suckers in the West. It sometimes reaches nearly 3 feet in length and weighs over 12 pounds. Once abundant, it is now rarely found in the Upper Basin.

INDEX

Adobe Lake, 86

All American Canal, 89

Allen, 15

Angels Window, 48

Antelope Island, 41

Arapaho National Forest, 10

Arapaho National Recreation Area, 10

Arches National Park, 25

Bard, 89

Barstow, 69

Big Gypsum Ledges, 61

Bill Williams River, 75

Black Canyon, 64

Black Mountains, 64

Blankenship Bend, 73

Blue River, 13

Blythe, 80, 83

Bonanza, 20

Bonelli Landing, 61

Bond, 13

Bookcliffs, 18

Boulder Basin, 63

Boulder City, 63

Bouse, 80

Bright Angel Creek, 48

Bright Angel Trail, 49

Bullhead City, 66

Bullfrog Marina, 39

Burns, 14

Callville Bay, 63

Cameron, 46

Canyonlands National Park, 29

Castleton, 24

Cataract Canyon, 38

Chemehuevi Valley, 73

Cibola, 86

Cibola Lake, 86

Cisco, 23

Clifton, 18

Coahuila, 92

Collbran, 17

Colorado National Monument, 18

Colorado River Indian Reservation, 80

Cottonwood Cove, 64

Courthouse Towers, 25

Cove Canyon, 38

Dangling Rope Marina, 40

Dark Canyon, 38

Davis Dam, 65

Davis Lake, 86

De Beque, 17

Dead Horse Point State Park, 28

Delmar Butte, 60

Derby Junction, 14

Desert View, 49

Detrital Wash, 61

Devil's Cove, 60

Devil's Elbow, 73

Devil's Garden, 25

Dewey, 24

Dirty Devil River, 38

Dolan Springs, 64

Dolores River, 24

Dotsero, 14

Eagle River, 14

Earp, 76, 80

East Rim Drive, 49

Echo Bay, 62

Ehrenberg, 80

El Dorado Mountains, 64

English Village, 74

Escalante River, 40

Estes Park, 10

Farley Canyon, 39

Ferguson Lake, 86

Fish Ford, 23

Fisher Towers, 24

Fishers Landing, 86

Flagstaff, 42

Fort Mohave, 66

Fredonia, 46

Fruita, 20

Gila River, 89

Glade Park, 18

Glen Canyon National Recreation Area, 38

Glenwood Canyon, 15

Glenwood Springs, 15

Golfo de Santa Clara, 92

Good Hope Bay, 39

Gooseneck, 28

Gore Canyon, 13

Granby, 12

Granby Lake, 10

Grand Canyon Village, 49

Grand Junction, 18

Grand Lake, 10

Grand Mesa, 17

Grandview Point, 49

Gregg's Hideout, 60

Green Mountain Reservoir, 13

Grizzly Creek, 15

Guardian Peak, 61

Gunnison River, 18

Gypsum, 14

Gypsum Point, 38

Hall's Crossing Marina, 39

Havasu, 73

Havasu Falls, 55

Headgate Rock Dam, 76

Hemingway Harbor, 63

Hermit's Rest, 49

Highline Lake, 20

Hite Marina, 38

Hittle Bottom, 24

Hole in the Rock, 40

Hoover Dam, 63

Hopi Point, 49

Horsethief Canyon, 20

Hot Sulphur Springs, 12

Hualapi Trail, 55

Imperial Dam, 89

Imperial National Wildlife Refuge, 86

Imperial Reservoir, 89

Inner Rim, 50

Islands in the Sky, 29

Jacob Lake, 46

Jumbo Pass Road, 60

Kaibab, 48

Kanab, 41

Kanab Creek, 55

Katherine Landing, 65

Kingman, 55

Kingman Wash, 63

Klondike Bluffs, 25

Kremmling, 13

La Sal Mountains, 24

Laguna Dam, 89

Lake Havasu City, 74

Lake Havasu State Park, 75

Lake Mead, 60

Lake Mohave, 64

Lake Moovalya, 76

Lakeshore Road, 63

Lakeside Mine Road, 61

Las Vegas Wash, 63

Last Chance Bay, 41

Lathrop Canyon, 29

Laughlin, 66

Lee's Ferry, 42

Little Colorado River, 46

Lipan Point, 49

Loma, 20

London Bridge, 74

Loop (The), 29

Mack, 20

Marble Canyon, 42

Martinez Lake, 86

Mather Point, 49

Maze, 29

McCoy, 13

Mead, 60

Mexicali, 92

Midland, 80

Mittry Lake, 89

Moab, 28

Mohave, 64

Mohave Point, 49

Moki Canyon, 39

Monarch Lake, 10

Moovalya, 76

Moran Point, 49

Morelos Dam, 92

Muddy River, 62

Murgula, 92

Nankoweap Canyon, 46

Natural Bridges National Monument, 39

Navajo Bridge, 42

Navajo Canyon, 41

Needles, 69

Needles (The), 29

Nelson, 64

New Castle, 17

North Rim, 48

Northshore Road, 62

Oatman, 69

Orchard Mesa, 18

Overton, 62

Oxbow Lake, 83

Padre Bay, 41

Page, 41

Palisade, 17

Palo Verde, 83

Palo Verde Diversion Dam, 80

Palo Verde Lagoon, 83

Paria River, 42

Park Moabi, 70

Parker, 76

Parker Dam, 75

Parker Dam Road, 76

Parshal, 12

Peach Springs, 55

Pearce Ferry, 60

Phantom Ranch, 48

Picacho State Recreation Area, 86

Pigeon Canyon Road, 60

Pima Point, 49

Pittsburg Point, 74

Point Imperial, 48

Point Sublime, 48

Potash, 28

Powell (Lake), 38

Pumphouse Recreation Area, 13

Quartzsite, 80

Rainbow Bridge, 40
Rancho del Rio 13
Rangely, 20
Red Canyon, 39
Rifle Gap Reservoir, 17
Rio Hardy, 92
Ripley, 83
Riviera, 66
Roaring Fork River, 15
Rock Creek Bay, 41
Rocky Mountain National Park, 10
Rogers Spring, 62
Rose Ranch, 23
Ruby Canyon, 20

San Juan River, 40
San Luis, 92
Sea of Cortez, 92
Searchlight, 64
Senator Wash Reservoir, 89
Shadow Mountain Lake, 10
Shoshone, 15
Shoshone Point, 49
Silt, 17
Slickrock Bike Trail, 25
South Cove, 60
South Kaibab Trail, 49
South Rim, 49
State Bridge, 13
Steamboat Springs, 13
Supai, 55

Temple Basin, 60
Thompson Bay, 74
Topocoba Trail, 55
Topock Gorge, 70

Topock Marsh, 70
Toroweap Point, 55
Trachyte Canyon, 39
Troublesome, 12
Tusayan, 49
Tuweep, 55
Twin Bridges, 14

Unknown Bottom, 29

Valley of Fire State Park, 62
Vasey's Paradise, 46
Vega Reservoir, 17
Virgin Basin, 61
Virgin Canyon, 60

Wahweap Bay & Marina, 41
Warm Creek Bay, 41
Webb, Del (Properties), 39
West Rim Drive, 49
Westwater Canyon, 23
White Canyon, 39
Whitewater, 18
Williams Fork Reservoir, 12
Willow Beach, 64
Willow Creek Reservoir, 10
Windows Section, 25
Windsor Beach, 74
Windy Gap, 10
Winter Park, 10
Winterhaven, 89
Wolcott, 14

Yak Point, 49
Yellowstone National Park, 12
Yuma, 89

ORDER FORM

"Recreation on the Colorado River"

FIRST EDITION

SEND TO: Sail Sales Publishing
P.O. Box 1028
Aptos, CA 95001

$9.95 Book
.65 Tax
1.25 Postage & Handling
$11.85 CHECK ENCLOSED

NAME: _____

ADDRESS: _____

_____

■■

ORDER FORM

"Recreation on the Colorado River"

FIRST EDITION

SEND TO: Sail Sales Publishing
P.O. Box 1028
Aptos, CA 95001

$9.95 Book
.65 Tax
1.25 Postage & Handling
$11.85 CHECKS ENCLOSED

NAME: _____

ADDRESS: _____

_____

NOTES

take only
memories

leave only
footprints